Flying In, Walking Out

Flying In,
Walking Out

Memories of War and Escape
1939–1945

Edward Sniders

LEO COOPER

First published in Great Britain in 1999 by
Leo Cooper
an imprint of Pen & Sword Books,
47 Church Street,
Barnsley, S. Yorkshire, S70 2AS

ISBN 0 85052 693 0

A CIP record for this book is available from the British
Library

Typeset in 12/14½pt Original Garamond by
Phoenix Typesetting, Ilkley, West Yorkshire.

Printed in England by
Redwood Books, Trowbridge, Wilts.

Contents

To Flight Lieutenant W.A.P. Manser
Royal Air Force.

A Hurricane Pilot of 134 Squadron RAF
and my good friend and comrade in our
two escapes.

And to my Father, Alfred Sniders, who
served with The Royal East Kent
Regiment, 3rd Foot, 'The Buffs' in the
First World War and with the 58th Rajputs
on the North-West Frontier of India.

To Mary

What is the mystery of your loveliness?
Mystery wondered, darling, until greeting
Loveliness new moon-risen on your face
Shining through pearl mists at our dreamland meeting
In our aerial sleep-known trysting place.

What is the mystery of your loveliness?
Is it your cheeks kissed by the English sun
Or those imagined lips with the hay caress
Of their shining secret from everyone
Is it your hair in a sea wave sweeping
Green-gold from your brow like some lovely dame's
In ancient tapestry with twilight creeping
Through faded forests penned with sunset flames?

And if slow sorrow of awake weeping
From dreams of you always to this hopelessness,
What healing joy the daily fresh sunrise –
What is the mystery of your loveliness?
Your voice heard yet, or the tenderness
Of dream-joy in your dream-blue eyes?

ES
Germany, March, 1944

Acknowledgements

This book was begun some years ago when Christopher MacLehose at a dinner suggested I write it. Ultimately, and with the encouragement of many friends and the immense help of my editors and my publisher, this story has sailed happily into Pen & Sword.

I have now the pleasant task of recording my expressions of appreciation to those who have encouraged and assisted me. Many friends were good enough to read earlier versions of the book and provide comments. To all I am grateful. I should particularly mention Colonel E. Bearby Wilson, Founder and Commandant of the Abu Dhabi Defence Force, Andrew Best, John Cannock of Lima, Peru, Lord Charles and Lady Cecil, Lord Chalfont, Nancy Dargel of the Oxford Society, Peter Fullerton of Magdalen College, William A.P. Manser my companion in escaping, Sir Michael and Lady Marshall, Count and Countess Michalowski, Walter Noel in Greenwich, Connecticut, Ivan Prinsep in Montreux, Switzerland, Major General Sir Roy Redgrave, Lord Schiemann, Countess F. von Schönborn in Geneva, The Earl St Aldwyn, Patrice de Vallée in Paris, and George Waterman in New York. I am most grateful to them for their support and understanding and for inspiring much of the enthusiasm that went into the creation of this book.

I am indebted to the Imperial War Museum, especially Philip Dutton of the photographic library; to the staff of the London Library who have sent their books from London to Geneva; to Irène de Charrière for typing the manuscript time and time again and also to Karen Stylianoudis.

To my publishers in particular Brigadier Henry Wilson and to my editor, Brigadier Bryan Watkins, I express my warmest thanks for their highly expert editorial suggestions and professional improvement.

I would like to end with particular thanks to Brigitte de Saussure to whom I am deeply grateful for her patience and work.

Geneva
June 1999   E.S.

List of Illustrations

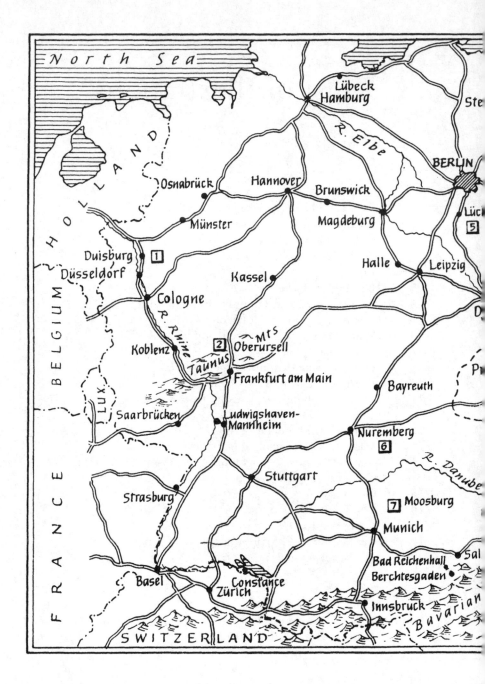

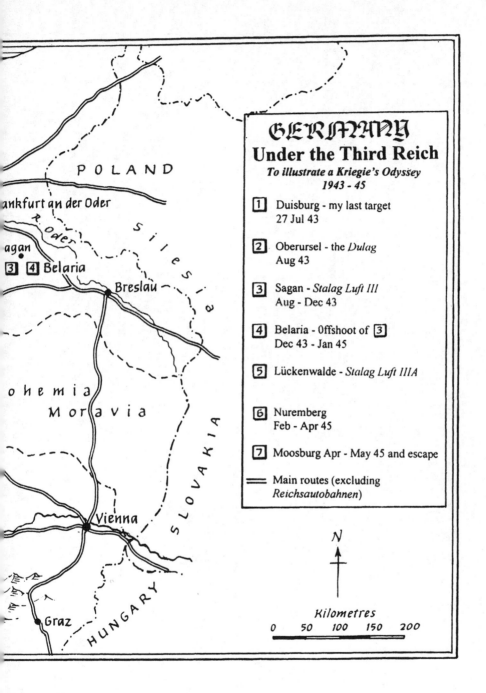

GERMANY
Under the Third Reich
To illustrate a Kriegie's Odyssey
1943 - 45

1   Duisburg - my last target
27 Jul 43

2   Oberursel - the *Dulag*
Aug 43

3   Sagan - *Stalag Luft III*
Aug - Dec 43

4   Belaria - Offshoot of 3
Dec 43 - Jan 45

5   Lückenwalde - *Stalag Luft IIIA*

6   Nuremberg
Feb - Apr 45

7   Moosburg Apr - May 45 and escape

===   Main routes (excluding
Reichsautobahnen)

N

Kilometres

0   50   100   150   200

POLAND

ankfurt an der Oder

R. Oder

agan

3 4 Belaria

Breslau

S i l e s i a

ohemia

Moravia

SLOVAKIA

Vienna

HUNGARY

Graz

Preface

'We might as well take them into the courtyard and shoot them now,' the German sergeant said to half a dozen field-grey troops who stood covering us with automatic weapons as we sat in chairs in a room of grey stone. 'It's got to be done soon anyway.' One of the soldiers grunted what seemed to me approval. The others went on looking inertly at us. My heart started to beat fast.

We were six RAF aircrew, two officers, a Squadron Leader and myself, a Flying Officer – both pilots – and four young sergeants, navigators and gunners. We had been shot down, each from a different aircraft, date and place. A Captain and detachment of *Feldgendarmerie* and a Gestapo civilian had arrested us an hour before in Charleville in northern France.

A potentially awkward circumstance was that we were all in civilian clothes, and carrying false identity papers.

My companions had not understood a word of what the sergeant said. I decided to leave it that way. I had some cigarettes in my pocket, lit one and inhaled hard, and as my heart continued to pound I concentrated on praying for help to die with dignity.

❖   ❖   ❖

People who were not alive at the time have asked what was the use of escape when practically nobody got home and, albeit infrequently, somebody got killed. The answer – though at the time few, if any, of the escapers perceived this – is that we were responsible for two considerable military factors, one of prejudice to the enemy, the other of benefit to multitudes of prisoners of war.

First, every time a 'Kriegie'* escaped from a camp, the police, the railway authorities and the home guards were alerted on a cantonal scale. When half a dozen or more got away there was a national alarm, and many thousands of Germans were deflected into trying to recapture us. The trickle of escapers, scattered through various camps and times, amounted to a continuous escape effort. This imposed a significant burden on the Germans, military and civilian. Many troops had to be stationed as guards in the camps although they were increasingly needed on the fighting fronts. The more 'escape minded' the camps were, the more was that the case.

This nagging waste of German effort was a sort of sucking sabotage. It was the remaining blow which the prisoners, as a force, could still deal to help the British war effort. This factor was known by some senior officers as 'Camp as a Military Operation'.

The second factor was that of morale. It was of immense benefit to a number of Kriegies within the camps. It was only too easy for them to suffer homesickness. Some of their sparse letters told of a brother killed in action, some, but more rarely, of a wife who had wandered away. Other prisoners suffered depressive self-reproach, rarely but terrible: if they had done this instead of that, they might still by flying in the squadron and their crews not dead. Amateur

* Abbreviation of the German *'Kriegsgefangener'*, meaning prisoner of war.

theatre, studies to equip oneself better for after the war, sports and gymnastics, bridge and so on could help. But, as Causely* asked:

> 'And why have you brought me
> Children's toys?'

All too often, there was a tendency to feel isolated among so many others in a present that offered nothing serious to do but wait for the long term future.

Hence the prime value of escape support – the clandestine factories working on an industrial scale in RAF camps with dozens of men, sometimes two hundred or more, working permanently at full stretch if a big tunnel was anywhere near its breakout, with special peacetime skills such as tailoring, to convert uniforms to civilian clothes, copying maps, forging identity papers, undertaking interminable lookout shifts, or disposing of sand – the vast list of clandestine activities under the aegis of the Escape Committees, and all this for the sake of the few who wished to try their luck.

These supporters were not going out themselves, but felt deeply engaged in the intended adventure, an activity not remote like the end of the war but excitingly near at hand. If the Germans discovered it prematurely, there was soon another. And all the time the community of helpers were pulling together for the good, continuously receiving and obeying instructions as they had before they had been shot down. Morale was raised and held, submission to discipline renewed, confidence returned. They could face and deal better with what the future might bring.

* The poet Charles Causley, who lives in Cornwall, is Fellow of the Royal Society of Literature and holder of the Queen's Gold Medal for poetry. He served six years in the Royal Navy.

And the catalysts for both the prejudice to the enemy and the upholding of morale were that only a few hundred of us in the various camps throughout the war wanted to get out and try our luck. I maintain that the question 'What was the use of it all?' is amply answered above.

What remains to be answered is 'Why did they do it?'

Our clique and those like us did not worry about other people's morale. The 'Camp as a Military Operation' had not filtered down to our level. If it had, we would have spurned it as a by-consequence, never part of our purpose, our grail, the thousand-to-one chance of winning the lottery. We were as aware as anyone of the probabilities, the legendary one in fifty times fifty and the logical possibility of a bullet. Why were we wasting our time, as the friendly and highly helpful majority may have thought, if too polite to say so?

It is to be remembered that air crews like ourselves had been living under a specific reality, namely that the tour of operations on the squadron in 1943 was thirty raids, after which we were taken off operations, and the number of those who reached the end of a tour was unpromising. The only way to deal with that reality was to shove it behind one altogether, which we did in favour of other realities, no less real. We loved flying, our operations were exciting to the highest degree, we were bound by an intense loyalty to our squadron.

Let me quote my own, and much admired Flight Commander John Bergrrens, from Michael Bowyer's *2 Group RAF*.*

Morale was very high on the Mosquito Wing and competition to have a place on the various raids was always very keen, despite the fact that losses were fairly high and almost

* (Faber & Faber).

inevitably one or two aircraft failed to return from operations. And later: These were memorable, exciting days. I don't think that any member of the Wing – be they air or ground crew – would have traded his place for any other job in the Royal Air Force at that time.

And so it was too with the hard core of escapers. We simply shoved the statistics behind us and favoured instead a bundle of other realities, weighing differently with different individuals – youth, homesickness, anger with the goons,* boredom, yearning for the squadron again, almost anything, even the Kings Regulations; and, identically for every one of us, the thrill of getting out, the tang of danger (the equivalence of extreme rock climbing), a hope of home next month: the Swedish sailor who hides you in his ship for a thousand pounds reward from the British Consul in Stockholm, the night crossing of the frontier into Switzerland after travelling through Germany.

The only real answer to the question is, we tried and tried to escape simply because we could not bear not to.

---

* Kriegie slang for the German prison guards. Goon box – a guard tower.

1

September, 1939,
to December 1942

I suppose the road – of which this looked like being the end – started with books about Biggles* when I was a schoolboy so that, as boys dream, I thought I would be a fighter pilot if there was a war when I grew up.

I was, I thought, grown up when Great Britain and France declared war on Germany on 3 September, 1939. I was in Texas at the end of a summer holiday after my last term at the City of London School, and going up to Magdalen College for my first term at Oxford. Instead of sailing back to England, the judicious course seemed to me to go to Canada and join the Royal Canadian Air Force. But, fatally for this policy, I visited the British Consul General in Chicago on the way. He asked me what I thought I would do if I was turned down in Canada as I couldn't get any more money from home because of the currency prohibition now in place. I hadn't thought of that and, while I was still under the shock, he booked me on to the SS *United States*, bound for England, as the *France*, on which I was booked to return, had been commandeered as a troop ship.

* Stories by Captain W.E. Johns about air-fighting in the first World War which delighted the young boys of Great Britain until the Second.

About 400 other young people and I had a cheerful, dancing crossing, wrongly alarmed by a whale or two. We little knew that fifty-three Allied ships had been sunk in September for only two U-boats. For us the first real sign of war was the cluster of barrage balloons floating, grim and grey, over Southampton as we docked. I went to the Royal Air Force recruiting office nearest to my mother's flat in London to be told: 'Go away and come again in another day'.

So it was Magdalen after all, for three terms made fascinating by my tutor, C.S. Lewis, and then at last acceptance as a trainee for air crew duty. By that time Dunkirk was over, and Great Britain stood alone, with some brave Free French, against the German Reich, now bloated with occupied Europe, and with Stalin as their accomplice.

Basic military training for me began at Blackpool. A long Cadet training followed at Torquay, where our parades and marches sought to emulate the Guards, and the classroom dealt with the mathematics of becoming air crew. One day, in over enthusiastic gymnastics, I dislocated my right shoulder so badly that Watson-Jones, chief orthopaedic surgeon for the Royal Air Force, put it right with an operation he had invented, all of which put me back from getting my wings by at least four months, to everybody's exasperation, and especially mine.

Then came the high enjoyment of flying for the first time on 13 May, 1941, at Elementary Flying School at Stoke-on-Trent. For two months we flew wonderful little Miles Magisters, monoplanes, with our heads in the air behind the windscreens, and lots of aerobatics. It was at this time that Mary Salter and I got married. We had been friends from my first term at Oxford. Because we were both slightly under twenty-one, we did so with our parents' permission.

From Stoke I was posted to Kidlington to fly Oxfords, solid planes, designed for training, their twin engines meaning that I would be in Bomber Command. By good luck

7

the station was an easy bicycle ride from Hampden House, where Mary's parents lived, and the Commanding Officer allowed me to sleep there during the three months of training, instead of in the camp. It was a house of some charm and the scene of many happy parties for fellow cadets and some of our instructor pilots. In the nineteenth century it was immortalized by the painting 'Where did you last see your father?' and its thousands of prints throughout the country – the brave little boy in the Civil War confronting the wicked Puritans.

The house also had to endure a short modern drama one night. Half way through the pilots' course, we were practising night flying near our airfield and a German intruder shot down one of our planes in the dark. The gunfire and the plane crash were within easy hearing of Mary and her parents, who had to wait hours until early dawn, hoping that I would return.

Another month, and we graduated on 3 September, 1941. The longed-for wings were pinned on our battledress, transforming cadets into grown-up aircrew – pilot officers or flight sergeants. Then we scattered, proud and joyful, on a ten-day leave during which we newly designated officers had to have our dress uniforms made.

Meanwhile, hidden powers were deciding at which Operational Training Unit (OTU) we would be trained to fly and to fight, in aircraft of types recently withdrawn from operational service. After some three months of that training, we would join our allotted squadrons. There, after a further brief period of training to convert us to the squadron's aircraft, we would be allowed to fight the enemy ourselves.

* * *

At the end of September, I was posted to Number 42 OTU (No 2 School, Army Co-Operation), at Andover, along with only two or three others of my Kidlington companions. The other members of the class were at least three or four years our seniors. They were army officers from various regiments, some of whom had been commissioned before the war. All had fought in France and returned from Dunkirk, over a year earlier. Seeing no prospect of further action for years ahead in the Army they had requested, and received, transfer to the Royal Air Force. We, the cadets of a month ago, were a little awed by them and they mildly condescending to us to begin with. But we soon rubbed along quite well in the common enjoyment of flying our aeroplanes in the daytime and spending cheerful evenings in the mess. One memory I have is of an Irish ex-Captain O'Mallay who won a whole weeks' pay from me at poker. As the war proceeded, until my captivity, I read a sad number of their names in *The Times'* list of 'Killed in Action'.

Our planes were mark I Blenheims for the first week and then Mark IV Blenheims. Twin engined, normally with room for the pilot and a navigator, now an instructor and a pupil. They had been used as fighter bombers at the beginning of the war and sent to targets in Germany, where they were massacred because our fighters lacked the range to cover them. Subsequently they were employed against targets in Occupied France and the Netherlands, faring somewhat better because they had fighter cover at last. Then, to everyone's relief, they were superseded for action in the autumn of 1941 by the Bostons that we received from the United States, and then our own de Havilland Mosquitos a year later.

Married officers were allowed to live with their families in the town at their own expense if they wished, as Mary and I did. We took rooms in the house of a former actress who was visited two or three times by Noel Coward, whom we found charming.

Wedded life now began. It included mutual reproaches for untidiness in our rather cramped rooms, but was otherwise agreeable, including Mary's cooking at which, I remember, she shone, particularly with duck and chicken.

After a few days training, my first flight as pilot on a Mark I Blenheim was on 1 October, 1941. Our principal activities were long distance flights, practice for the bomber attacks we would be making later, and aerobatics, hopefully, to avoid destruction by German fighters.

As to the aerobatics, after the sporting Magisters with which we had been happy looping the loop, putting into spins and coming safely out again, the Blenheims felt heavy and less inspiring (our Oxfords had not been used for this sort of fun). The manoeuvre which I remember best was to pull the stick back as suddenly and as hard as possible to jerk the plane vertically up and, as it slowed just before the point of stalling, to slam down into a vertical dive which, we were told, would make it difficult for an enemy fighter to follow us shooting. It was great entertainment and would have been a morale booster for the gullible, which included none of the pupils I knew.

Early in November, my future crew were allotted to me, two sergeants younger still than my own recent twenty-one, Sergeant Hinds, navigator, and Sergeant Dean, gunner. From then on we flew together to the end of our course.

Most of our flying was done over the wide plain of the South West to Devon and Dartmoor, and Cornwall's breathtaking cliffs, white-edged with foam. Occasionally, we flew far out over the grey Atlantic. We never flew above the English Channel alongside us, for it was the pathway for the Royal Air Force attacking targets in occupied Europe at that time and, perhaps, for German prowlers for whom we would have been easy meat.

Soon we grew able to emerge from cloud on a prescribed course and perceive our whereabouts at once, without looking at a map, from the harlequin pattern of landscapes below, the vast white horses and giants cut before history in the green sides of the downs, the mauve-tinged moors, the dolls' houses and cathedral spires of south-west England.

In all our activity we were children of the weather, climbing into sunshine, looking down through pale grey rain or darkened treetops, in storms with silent lightning flashes around us, sometimes between cumulus clouds which it might be fatal to enter, the theatre and abiding playground of gods born millions of years ago.

One bright morning, on 11 December, two weeks before the end of our OTU course – I can see it as clear as yesterday – a ring of happy faces, putting on our flight kit in front of the crew hut, we were thrilled by the news of the US declaration of war on Germany and Italy.

Another leave, another family parting, and I, along with Sergeants Hinds and Dean, joined our first squadron in February, 1942, 88 Squadron of Two Group, Bomber Command at Attlebridge in Norfolk, a year of training over. They were flying Bostons, strong, solid planes, faster than Blenheims. The pilot and navigator flew in the nose, and the gunner in a turret in the rear, affording a degree of protection for us against the German fighters.

The latter were in considerable use around the Squadron's principle employment, known as 'Circuses', bombing pin point targets from ten thousand feet – important railway yards, factories, particular enemy installations, and the like – in occupied territories, mainly in France and the Netherlands.

Another use was flying down on the sea, as low as possible to miss radar on the way to bomb German ships. They were heavily armed to fire at the Squadron's bellies in the last moments of encounter when our planes would zoom up from the water level to dive again at the proper angle, and with all speed to hurl bombs at the ship. Up and away above the level of the masts, and jinking as hard as hell to avoid the shells speeding at our tails.

11

I was initiated into the Boston by just over one hour's practice with the rightly nicknamed 'Baldy' Matthews, whom I had last seen at Kidlington and who had somehow got ahead of me.

Passed out as efficient by the Flight Commander, our crew did a few practice trips of about an hour at a time, practising formation and low-level bombing. We then, at last, joined the Squadron Leader for initiation in a formation flight of five or seven planes in a broad arrow, wing tip almost to wing tip it seemed.

It was the first time I had flown that way and after some seconds of tension I clicked into place with a feeling of belonging to the formation, stronger and safer than the mere sum of its parts. He led us well out over the North Sea, a slight mist on the sunshine, a pale grey and green on the waves down to which we dropped as soon as we left the coast. Then, still arrowed together, quick up and thrust down, and quick up and down again and again and again to get the idea of what would be needed, and to show that it would have to be perfectly done, never mind the guns which would be shooting up at us.

Next day we were at last ordered to see some action – an attack on a reported ship in the North Sea. All was ready, I had done the cockpit drill, we were a second before starting our first operation and, cursed luck, it was called off. We were ordered to cut off the engines, bomb crews came round and took the bombs off. We were up and off again for another dummy act away over the green, beckoning sea.

And so we did our graceful swoops down again, we tore along a couple of yards above the waves (on another, similar, occasion in the Squadron one of the boys came back with the body of a fair sized fish in his air vents), and then the quick climb and the murderous dive, and all in play.

When we crossed the coast again and reached our airfield, we broke formation and circled to land one by one. The wind had changed and instead of the long path we had used for take off we were ordered to land on a much shorter emergency path. This had a farm house near by, about halfway along it, which was due to be dismantled. As the path was short I, like the others, came low into land, slow with plenty of engine, 'hanging on the props' as it was called. At exactly the best point for a disaster, my port engine cut. At that slow speed I couldn't get over the house towards which the starboard engine was pushing us so I put the port wing down for a crash landing to miss it.

Crashed it was, and much more than I wished. I felt the port wing crumble, and then a fearful squeeze on my back. I shouted or thought I heard a shout, and without losing a moment of conscience I was looking at bright long grass less than a yard from my eyes, hanging by my straps. I found I could move nothing, I was pinned by wreckage, a slight trickle of bright blood was dribbling onto the green blades of grass.

The airfield had not been long in operation. But it had seemed long enough for its crash crew, two Red Cross boys who, with their ambulance, had been bored beyond imagination, sitting for ever behind a hedge, waiting and waiting for something to happen. Now at last it had. I heard the ambulance howl, my hips and back began to feel uncomfortable, and quickly more so, and I heard one say to the other, 'He's had it, poor sod': the words have stayed with me forever. They started swinging axes to cut the metal away, and I shouted, 'No, I bloody haven't and stop chopping or you'll cut my leg off'. Parts of my body were in unusual situations and the enthusiasm of the boys scared me, not to death, we'd just done with that, but enough for them to forgive my bad manners.

From the corner of one eye I saw that an engine, thrown well away from us, had flames coming from it. Miraculously, both of my crew were standing upright nearby. Then the Red

13

Cross dug a needle into me and everything seemed less disagreeable. I remember being carried away on a stretcher and receiving a pitiful look from a New Zealand Flight Commander whom I knew and liked. I tried to wave to show he should cheer up.

I thought the public emergency ward in Norwich hospital was unpleasant, but as I was being pumped with morphia I may have judged it ill. I asked for Hinds and Dean, but they had gone.

I kept dreaming awake that I was floating in the Channel on a one man rescue raft made of pain instead of rubber. Then Mary, my wife, was sitting next to me. I was half awake, I saw she was wearing a white tweed suit. We both managed to laugh but then the broken bones rubbed together and all was bitter pain. Mary had gone.

While I was still in the casualty clinic at Norwich, Hinds and Dean visited me and I realized again how much I liked them. They were killed a month later over Holland on a 'circus' along with my successor.

Soon I was moved to the RAF hospital at Ely and Watson-Jones again. It turned out that the Boston had broken my pelvis in several places, and other bones thereabouts. He put it all together again and told me that the X-rays had their place in his gallery of horrors.

In the same ward were a number of American volunteer fighter pilots of the famous Eagle Squadron who brought many German aircraft down before the United States joined the war. They were still recovering from severe wounds from the Battle of Britain, cheerful and tough, and from them I learned serious poker at last, not without pain on the way.

Another game was Battleships, more anodyne, played on paper in opposite beds, a game more fascinating than one might think. As we were in opposite beds across the corridor, Max Aitken was my opponent, also wounded, highly adroit, as in many greater things, a charming officer.

Then came convalescence for some weeks on the south-west coast in a vast, white Edwardian hotel near Torquay, converted to the Officers' Convalescent Station, high on a green hill looking over the cliffs and far into the Atlantic. There, every day, we spent hours on our specified exercises under the eyes, and sometimes the hands, of well briefed physio men in the gymnasium and swimming pool, getting us better and better.

The doctors included orthopaedic surgeons since many officers needed further repairs. They and the nurses – many of whom were young volunteers (the VADs) – were immensely efficient and kind. One in particular lives in memory, Australian, small, blonde and fairy-like, universally known as 'Tinker-Bell', a most sweet nature. We all loved her.

I shared a room with another pilot, Johnny Skinner. He had come to the Station a week or two before. His ailment was a broken ankle when he had to ditch his Boston in the Channel, it having been ruinously shot up by German flak on leaving the enemy coast after a 'circus'. Landing a plane in the sea was no picnic, but Johnny managed it well, there was no damage to his crew, and one of the air-sea rescue boats picked them up in the evening.

When deemed well enough by the doctors, we went out in the evenings, mostly to the other huge hotel lower on the sea front in Torquay itself. At that time it was mostly filled with the wives and daughters of successful businessmen who sent them from the bombing in London and joined them at week-ends, their sons probably in their boarding schools or doing the same things as us. The orchestra of the hotel provided dancing at night, the hotel invited all of us who were able to come, and good times were had by many.

A pastime of Mary's and mine was to spot the officer of each night most fitting Ernest Dowson's:

I cried for madder music and for stronger wine
But when the feast is finished and the lamps expire
Then falls thy shadow, Cynara! the night is thine;
And I am desolate and sick of an old passion,
Yea, hungry for the lips of my desire:
I have been faithful to thee, Cynara! in my fashion.

We were not only young, but also immature.

One candidate I remember still, well named Randy by us all when out of feminine hearing – a Canadian. He had broken a leg which still had metal round it, had just been allowed out from the Station, and danced like the devil, telling every girl who would listen a harrowing tale of heroism. We, his colleagues, knew but never told, it was too funny. In very truth he had broken it jumping out of a window when a husband inopportunely opened the door.

After several nights he danced a step too many and his precious leg snapped again in its cage and he collapsed into the arms of his fancy. Away, poor Randy, back to painkillers and a single bed, no girls to tell your truthless stories to, gone good times.

Johnnie was healed and returned to his squadron, and soon after I to mine. After a trial in a Boston on the ground to see if I remembered all the controls, which I did, I was sent on a practice flight. Up in the air I was dismayed and intensely embarrassed to find that sitting in the pilot's seat was painful far beyond a joke. I landed, our station doctor took one look at the situation and sent me back to Watson-Jones to relieve me of my coccyx. Nobody had noticed among all the other pieces that it had been broken and grown together badly.

That took another month or so, and then another visit to the same Convalescent Station at Torquay. Same but different: repairs were going on to scattered damage on some of the walls facing the sea, and men were at work on the

16

roofs. Indoors, great surprise: Johnnie Skinner crossed me in the entry hall and explained.

The big white building always had painted Red Cross identifications on the roof and it was certain the Germans knew it was a hospital. Indeed they must have known by now that British air crew were its inhabitants. Against all we expected from the *Luftwaffe*, three weeks ago, in the morning, a flight of seven German fighters had risen from the sea surface they had been hugging and, flying just above the cliff and straight at the Station, sprayed the building with their guns, over the roof and circled back, and released a few bombs at the hospital, of which three hit. Then they flew away, over their cowardly sea.

During this attack a popular orthopaedic surgeon (I remember he had a Russian name) continued an operation on an officer which he had started before the attack, the operating table was running with blood – his own.

Randy was fixed in his bed by the tension wires in his famous leg and shouted for the nurses near him to dive under him for more safety and soothed them down, so he was not all bad.

And the appalling horror; our sweet little Tinker-Bell lost both her hands and died some days later.

Johnnie explained his reappearance. Back at his squadron he had fought two more 'circuses', high over the Netherlands, and then a third over western France. After 'bombs away' that time, a German fighter came from the rear firing its guns. It missed, as did Johnnie's gunner from the turret, and swung away to come again. Johnnie plunged, the German followed, Johnnie was jinking, the bullets missed again, the German was getting nervous of Johnnie's dive. With a tremendous struggle Johnnie flattened near the ground pointing home, 'ware high tension cables, until he had shaken the fighter off.

The sea was just ahead. He kept low to avoid flak, and then

the hell of a crack, it was flak alright, a small flame was showing from the port engine. Johnnie cut it and skimmed west on one engine just above the waves to get no more flak. For a while the fire didn't seem to be spreading while he made headway towards the English coast. Then, all of a sudden, it began to spread and Johnnie had to ditch again, rougher this time but got away with it. Into their rescue rafts once more, and again the air/sea rescue boys found them in the evening. He had torn his other leg and lost some blood, and that was all for himself and his crew.

We were both glad when we came to leave the Station and its recent memories, I this time before Johnnie. The Medical Board passed me fit for flying but not yet for full operational flying, and I was posted to Norfolk, RAF Marham in December, as a staff pilot in 1655 Mosquito Training Unit.

Coastal Sortie

Close the formation and wing-tip to wing-tip,
Proud and secure like arm-linked friends,
Stern for north waters, deep in the gold mist,
Till long the sea's rim crouches the ship.
'Open out boys – going in now', the leader
Turns and we wheel and dive down to the spray,
Down close to the water swirling and hissing,
Splashed dully white, so soft for a sleep,
Wide like the Bedouin circle the tanker.
Four seconds part us. Now in to attack.
Bounding and plunging in strong, angry gallop,
With quick, skidding turns as we dodge through the flak,
Tracer springs from us to frighten the enemy,
Frighten and smash him and silence his guns.
Now we are on him, the ship huge before us,
Naked and helpless awaiting the blow.
We have no mercy, trophies of heaven,
Cold, cruel sabres, cut down our bombs.
Up, up we leap and our bodies jerk backwards,
Up, up and over the sharp reaching masts.
Then down again, down to the jagged grey water,
Writhing and bounding away in the mist.

E.S.

2

Sorties Over the Coast

Marham was the base for 105 and 139 (Jamaica) Squadrons, 2 Group, Bomber Command, flying Mosquitos. The Station Commander was Group Captain 'Digger' Kyle. The squadrons were commanded by Wing Commanders Hughie Edwards, VC 105 Squadron, and 'Reggie' Reynolds, DSO DFC 139 Squadron.

The Mosquitos were the fastest British aircraft in operation at the time, twin engined, constructed of laminated wood, powerful and sensitive. They had a crew of two, the pilot and the navigator, who doubled as the bomb-aimer.* The two squadrons were flying them in daylight low-level bombing operations against 'precision' targets such as marshalling yards, enemy ships and the Gestapo headquarters in Oslo on the anniversary celebration of the Quislings.

Sorties were flown almost daily and sometimes twice in the day. The Mosquitos would climb from Marham one behind the other, get into formation, descend to the East Anglian coast and then low out across the sea. Approaching the enemy coast and beach guns, they opened formation, flew flat out, jinking if fire was encountered, closing again, hugging the contours, often lower than the tree tops to the

* Bombardier (US).

target, jinking in the ground fire as they neared it. The navigator crouched over the bomb-sight in the nose of the cockpit. The plane lifted a moment and 'down' on the headphone, 'steady, steady' from the navigator to aim the bombs, and then 'bombs away'.

Frequently, a specific manoeuvre called 'low-level shallow dive' was used. Three Mosquitos, flying fifty feet or less from the ground, would drop their bombs on target with eleven-second fuses. At the same time, flying at 1500 feet, the other half of the attack would bomb the same target with instant explosion. It was essential for the first crew to be away just before the second. The combined explosions were devastating.

Then up over the target and sharp down as fast and low as could be to streak cross country to the sea and Marham – except for those hit and brought down slam into the ground and fire and death.

All officers at Marham messed together. Most of us were in our early twenties. Newcomers were frequent because of the high casualty rate at that time in the two squadrons. They were hospitably received in the mess, as were we, the pilots of 1655 Training Unit.

The faces of many young officers now come to my mind, businesslike in sober blue battle dress in the crew rooms and at the dispersal points where their aircraft and the ground crews waited for them, merry and talkative at dinner in the mess in paler blue uniforms with bright brass buttons.

So many young faces. Two most vivid.

Johnnie Skinner, whose arrival at Marham on transfer to 139 Squadron late in 1942 was a happy surprise. In the next weeks he flew, and told me about, six or seven Mosquito sorties. Then one more, near Christmas, taking off late in the afternoon. We heard that the others were back, but not yet Johnnie. Two friends and I went to the landing path, at night by then, and stood near a dimmed landing lamp in the dome of falling snow flakes which it made, and waited until the answer was plain.

One night in bed, a few weeks later, when I thought I was awake, I saw Johnnie standing on the other side of the room, lit as if by firelight, dressed in full flying kit, standing still, his face grave and sad.

And Tommy Thompson, Mosquito pilot, tall, sweet-natured, from his father's great sheep farm in New Zealand, who sang lovely, haunting Maori songs to his friends which we pressed him to sing over and over again with calls of: 'Do sing it again, Tommy. Tommy, don't stop singing.'

Requiescant in pace.

* * *

Before joining the squadrons, the crews arriving at Marham fresh from general training spent a short time in a Training Unit. There the pilots learned how to fly Mosquitos, with emphasis on low level flying, and the navigators gained practice in what they had been taught, especially when tearing along low, near the ground.

Every Mosquito manufactured was urgently needed for operations against the enemy. In the Training Unit we therefore used Bisleys for the pupil navigators, a variant of the Blenheims which were phased out of operations in mid-1942. In them we flew distances, day after day in all weathers but bad fog, over great stretches of England, Wales and parts of the sea, mostly at two to five thousand feet, and with plenty of low-level in specified areas to accustom our pupils to map reading in what would soon be their frequent medium. Each of these flights was about half as long as the average operational sortie but twice as frequent.

Flying the Bisleys with pupils was enjoyable work: beautiful green shires, mountains, moorland, coastlines huge below, the endless change of the cloud scenery around us, the heady excitement of low flying which, outside of operations, was rarely allowed. But of course nothing could compare with flying the Mosquitos on operations.

On 30 January, the whole of our 1655 Training Unit was closed down. That day was the tenth commemoration of Hitler's accession to power, and Goering was going to make a morning speech in Berlin to the German Air Ministry with much pomp, grandeur and glory. He was about to speak when three of our Mosquitos, led by Squadron Leader 'Reggie' Reynolds, tore across the city, each aircraft letting go a load of 4 × 500 lb. bombs near the spot, so that glory was changed to dashing under-cover.

By the middle of the day, the news was all over Radio Free Europe.

Then, at 4 pm Goebbels, Hitler's major speech-maker, was also to glorify the Führer. On the dot, at 20,000 feet above the city, three more Mosquitos unloaded their bombs and stopped it all, but Squadron Leader Darling and his navigator were lost.

Two months and some forty flights later, I was sure the doctors' idea of restricting me to non-operational flying was misplaced. And having lived with 105 and 139 Squadrons for that time I very much wanted to fly with them. So I applied to see Group Captain Kyle and formally requested the honour of joining one of his squadrons if I passed the medical board. He received me kindly and promised me a place.

Someone flew me to an airfield near the relevant RAF clinic. I had no trouble with the orthopaedic and other physical tests. Then I was surprised to hear that I also had to be seen by a psychiatrist, and in I went.

He was a dismal little fellow in the uniform of a Royal Air Force medical officer. Beyond a peradventure he wanted me to agree that I was being coerced into asking to return to operational flying, and banged away for at least ten minutes. When I was not to be persuaded, he wrote a line or two on a report sheet in front of him, and told me, in terms, to go and break my bloody neck any way I liked. This did not raise him in my esteem.

The medical board passed me fit for full operational flying and, after one more month and fifteen more instruction flights, for a total of fifty five, I was re-posted to the Training Unit, this time as a pupil for conversion to Mosquitos and operational training for daylight low level attack. I found the Mosquitos were a joy to handle as I had always heard – and so passed April.

Then came the news that the two squadrons had been transferred to 8 Group, Pathfinder Force, still under command of Group Captain Kyle. This meant that the daylight low level bombing sorties were over for the time being and we would now be flying high altitude at night in front of the thousand bomber raids and, on nights when they were not operating, on smaller raids of our own. It also entailed a specialized Mosquito night flying course during which Flying Officer George Hodder, a navigator on the course, agreed to join forces with me. In mid-June we at last reported as a crew for duty to 139 (Jamaica) Squadron.

I was lucky indeed to get George. He had already done one full tour of thirty operations, costly daylight attacks in Blenheims, mostly against German shipping. He was extremely efficient and well balanced, five or six years older than me, tough, cheerful and good company.

*   *   *

During the spring and summer of 1943 Bomber Command sent large forces of heavy bombers at night, mainly Lancasters and Halifaxes, to drop high explosive and incendiary bombs on German cities. Some, such as Cologne and Düsseldorff, we revisited several times in a month. Others, like Hamburg, were saturated on several nights in succession. It was there, of course, that a firestorm was developed and maintained, so that more than three-quarters of its buildings were destroyed and some 30,000 civilians killed.

The first of the heavies would take off in the early twilight from their bases in eastern England, and successive waves would maintain the stream. As they neared the target, aircraft of Pathfinder force, starting later to synchronize with them, and flying ten to fifteen thousand feet above them, opened the night's performance by dropping marker flares and incendiaries. The first heavies bombed on the starting fires, and their followers continued bombing onto the spreading conflagration. Seen from our altitude, the wild furnaces in the ruins shrank to patches of orange and gold in a black bed, like hot coals in a fireplace.

When the first planes closed with the target, hundreds of anti-aircraft guns would open and maintain fire. The even blackness, through which the air crews had been flying for hours, was suddenly the background for an immense fire-work display, below, around, above, with shells bursting in flowers of fire, shining tracer spouting up at the crews, search-lights swaying and sweeping across the night and, when a plane was caught in one of their beams, interlocking to hold it for the German anti-aircraft gunners. Here and there a touch of red would suddenly appear to us against the black, float for a moment like a tiny hanging lamp and then become a burning bomber which crumbled in slow motion, dribbling embers down.

Tens of thousands of British air crew were killed in this campaign against the cities. Over six hundred thousand civilians, of whom far more were women and children than men, were killed by the bombing or died in the infernos. The avowed purpose of the campaign was to destroy civilian morale. In this, it failed, as did the previous German bombing of Britain.

*   *   *

Inside that strategic picture, for George and I our first operational assignment in the squadron was to bomb Berlin from 25,000 feet on the night of 19 June, 1943.

A dozen crews gathered in the small lecture room where Group Captain Kyle gave us our general target briefing, and the navigators went off to draw their flight plans according to particular instructions which each of us had received as to altitude, tracks, and time of arrival on target.

The sun was setting when we carried our Mae West life-jackets and parachutes from the store, and the clear blues and pinks of a June sky deepened as the truck drove us around the airfield to the diverse dispersal points where our Mosquitos waited with their ground crews.

George and I climbed up through the trap in the under-belly of our plane and strapped ourselves into our seats. The entry/exit hatch of Mosquitos was not wide, and pilots climbing into it pushed their parachutes ahead of them rather than wriggling through with them attached behind. I started the engines and performed the prescribed cockpit drill, checking the rudder, aileron and flap controls, and feathering and unfeathering the airscrew blades, while the sunset colours grew darker outside. Then, everything being in order, we taxied along the dispersal paths in allotted sequence with the others, turned on to the runway which was now pointing to a shadowy horizon drained of colour, opened the throttles and sped into the invariable exhilaration of take-off.

The cockpit lights were always extinguished on night operations, the better to see outside and the less to attract enemy night fighters. Accordingly our only lights were the luminous dials of flying instruments, a minimal torch for George's charts, the Gee box that would guide us to our target, and the bomb-sight in the nose of the plane when the time came.

With our two thousand pounds of high-explosive bombs, we climbed on course for the enemy coast. Norfolk spread below us engulfed in the war-time black-out. As we gained altitude, the twilight reversed to sunset, the higher we flew the pinker it grew around us, the blue more vivid above. When we flattened into our allotted cruising altitude we were flying in a luminous blue dome. Our port wing tip was pointing north to the horizon, glowing orange from the northern sunshine. Below us, the North Sea lay dark and lost.

Suddenly, moonlight silvered the huge sea. The black shore of Europe lay stretched from port to starboard ahead of us, the enemy coast which we soon crossed.

On the intercom, soon after take off, George had begun the navigator's litany of instructions to me about height and air speed and courses with which he normally guided our sorties until the start of the bombing run up to the target. He would then crouch in the perspex nose of the plane looking through the bomb sight and calling for micro-changes of direction with: 'left . . . left . . . steady . . . right . . . steady . . . steady . . . steady . . . ,' and then 'bombs away'.

At that point I would do a tight turn as steep as I liked and head roughly east for England and Saint George, weaving the plane to counter any predictor shellfire bursting near us, diving to get out of searchlights when we were being coned by them. If I threw the plane about too violently for the compass, we would pick up a rough course east by jabbing the starboard wing tip towards the slight northern glow still discernible on the horizon at that time of year. Then we would tear away until we were out of trouble, the compass settled down, and George could resume his litany to bring us home.

But this first time, our sortie for Berlin, ended differently. An hour after we had crossed the enemy coast the port engine started to fail. The boost control, the engine revolutions control and the supercharger all went out of action. Berlin was still far away and we decided to abandon the operation. It was a highly unpleasant decision to take.

After re-crossing the enemy coast, I ordered the bombs to be jettisoned in the sea. But George found the bomb release mechanism would not work so, when we got back to our base, we had to land with a full load of unfused bombs. I could not depend on an engined approach as the port engine could fail at any moment, so we came in high and I cut back the engines to glide down to the smoothest landing I ever made. George and I climbed out and were driven to the de-briefing room to make our report, disappointed and extremely annoyed.

Two nights later, with no trouble at all, we flew to, and bombed, Cologne visually from 20,000 feet in full moon-light. So began a rhythm of ten more raids, adequately described in the entries which followed the Berlin fiasco in my official logbook – three times each on Hamburg, Cologne and Duisburg, and a raid on Munich.

In the small hours of 27 July, George and I landed back from the tenth of our sorties, our third to Hamburg. After de-briefing, eggs, bacon and bed we were told that we could go on leave that day for a week.

But the next entry in my logbook is for the same day, July 27, in the evening. It shows Duisburg as the target, Squadron Leader Price as my navigator, and 'missing' as our fate.

3

Reported Missing

Missing indeed we were.

After George and I had slept off our trip to Hamburg I told him that I would like to clock up one more raid before going on leave. George would hear nothing of this. 'Mrs Hodder's little boy is going on leave right now' was what he said, and promptly did.

So I talked about it to Reggie Reynolds, our Squadron Commander, and he fitted me into the Battle Order for the same evening with another navigator, Squadron Leader Price, an officer with plenty of operations to his credit. He was serving for the time being at Wing Headquarters. When I met him in the evening in the briefing room he told me he was working on ways of improving night navigation equipment and had put in a request to navigate a night Mosquito raid the better to study our current equipment in action. I could not have wished for a better navigator to stand in for George.

Our target was Duisburg. Our allotted aircraft for the night was Mosquito 'J'. I had flown her on three previous operations and had logged her performance as 'very good' after each. During these operations she had been hit by flak splinters in the port wing over Cologne and in the port engine nacelle over Munich. She was now back in service. I was well content to have her again, and looked forward to showing her performance to Price.

It was a misty night when we took off and we were at once in heavy cloud. Watching the dials of the instrument panel, we climbed steadily through impenetrable dark to around 15,000 feet, where we rose from the cloud top. There was no moon to be seen and our engines were sharp black silhouettes against a ghostly less-than-blackness.

The night was clear about us with no horizon. Above was medium grey, and it merged down to black at eye level. No stars were to be seen. At our allotted altitude of 28,000 feet we settled into our cruising speed across the invisible sea and the enemy coast, and along our courses towards the Ruhr.

From time to time, Price gave me a prescribed change of course and in the glimmers of his Gee box he seemed to be noting a series of readings. I flew straight and level, checking in turn the readings of the flying instruments and swivelling to search the featureless space for enemy fighters, my eyes focused to infinity.

It was a different trouble which hit us. With some ten minutes to go to Duisburg, the port airscrew 'ran away', that is to say, from one moment to another it became disconnected from the engine so that the slipstream blew it whirling viciously around its axis and, acting as an air break, it began to make us lose height, gently and steadily.

In a crowded couple of seconds, I cut the port engine, stopped its petrol, opened the starboard throttle wide, trod hard on the rudder control to keep us straight on course, reset the trim and stabbed at the button to 'feather' the airscrew as in the cockpit drill before take off. But this time the feathering mechanism refused to work. The airscrew continued to windmill and we, slowly, to sink.

As pilot it was for me to decide whether to turn back at once or not, and I said to Price that as we had come this far we should go on with one engine and bomb the target.

By the time we got there the aircraft had begun to judder. We had lost a few thousand feet, and at this lower altitude

30

anti-aircraft fire was coming up densely at us. There was nothing to be seen of the target in the dark below so on Price's calculation of our position we made the usual steady run up with shells bursting around us. We dropped our bombs and turned about, but without my normal dive for extra speed: we needed all the height we could keep to get back to the English coast before we ran out of altitude. And if we couldn't do that, I intended to make a night ditching in the sea, shining our headlights to judge the surface of the water, and hoping that the Air Sea Rescue Service would pick us up the next day in our escape rafts.

However, some ten minutes after we left the target area, and still lower, small flames began darting out of the port engine nacelle. I pushed the button for the fire extinguisher in the engine, the flames stopped, and we continued westward, hoping we might hold height better as we got into denser air. Then the flames began again, and this time the extinguisher didn't respond to the button.

I wondered if the juddering had cracked an oil pipe or a flak splinter had hit us where it hurt, and then that didn't matter as the flames took hold and grew brighter and denser and bigger and it wasn't going to be long before the petrol tank in the wing blew up, and us with it. I sat keeping the plane on course as Price clipped on his navigator's breast-type parachute and moved down in the dark to the hatch in the floor. The air suddenly rushing in told me he had gone. I sat and looked at the flames for a forlorn moment, undid my safety straps and addressed the matter of getting out myself.

I had not heard of any pilot parachuting out of a Mosquito. The profuse casualties of 105 and 139 Squadrons had been near the ground, and fatal, until our squadron's recent transfer to high altitude night flying. And nothing was yet known about the few crews which had not returned from our present night operations, save one officer named Clark. He was heard in a distress call as he sought to ditch his Mosquito at night on the way home and was shouting that he 'couldn't see the sea', he 'couldn't see the fucking sea' when the end came.

As for me, by the time I got to the exit the plane was tilting into a dive. I jabbed my legs out into the slipstream and had to struggle hard to follow: the parachute pack, harnessed to my bottom, all but jammed in the hatch. As the plane and I finally parted company I grabbed the release ring to pull it once my falling body had slowed to terminal velocity from the far higher speed of the plane at that moment. The next thing I remember was my head reeling in complete blackness, my body generally shocked and banged about, the grope overhead to find if the invisible parachute had opened, and the relief at touching taut cords.

I looked down into darkness, absolute except one flake of light, the burning plane still on its way to the ground, and a flash and brighter burning when it crashed. I guessed that Price and I must have baled out at around 10,000 feet. He of course would be landing miles away. I could but wish him well and wonder what would happen next.

What happened next was nothing for a good few minutes by my slightly luminous watch, and nothing to be seen, not even my arm, nor heard, but silence as I drifted down. Then a lovely thing happened, from one moment to the next the darkness was full of the smell of trees and grass, the sweet scent of countryside at night where there had been none before.

This also meant that the ground and I were about to meet soon, heaven knew how, an open field, a farm machine, a tree top to catch me, unpredictable before the shock in this darkness at the speed of a jump from at least a six foot wall, and the drift of whatever breeze might be blowing, unfelt with my parachute sailing on it.

Then suddenly my boots hit what felt like sand. I stumbled forward a pace or two, did not fall, and stood still. I felt the parachute settle all around me. I could still see nothing. It was time to get moving. To begin with, I dragged my parachute together and wrapped the Mae West around it. I hooked my left arm over the bundle. At once my right hand touched what seemed to be sheaves of a cereal in a loose shoot. I stuffed the bundle in it to hide it from the morning to come. In a pocket of my battle dress I had the escape kit which we were issued to help us if shot down: sweets to suck, an empty rubber bottle for water, bank notes corresponding to the countries we were flying over (Dutch and Belgian mine turned out to be) and a small luminous compass. On this I set a southwest course for Spain and started to walk, peering into the dark.

The fog in which I now groped had in a sense been developing over the last two weeks. Fifteen days ago, George and I had bombed Munich and flown back in full moonlight in the small hours. The port wing tip was pointing south at the long white Alps, brilliant against the night, and I remembered happy skiing holidays. A week later we bombed Duisburg in the Ruhr, and in my log book reported slight mist on return. Two days afterwards, on 26 July, it was Hamburg and I logged poor visibility. Now a bare twenty-four hours later, Duisburg again. I was stuck on enemy ground and it was Stygian.

4

Fugitive in Holland

I was evidently in a night ground mist: hanging from the parachute I had been able to see my aircraft burn all the way down in the clear blackness and watch its funeral pyre, and now I could see nothing at all.

I had to assume it was still West German countryside around me. I was in RAF battle dress, embroidered pilot's wings above my heart. My immediate purpose was to remove myself from my bundled parachute, to make what distance was possible in my chosen direction until dawn, and then to find a hiding place to rest and pass the day. The vital point was for no one to see me until I could get out of German territory, if I was still in it, and into Holland where I could hope for help. Evasion was now the name of the game.

In a few minutes the mist was thinner, or my eyes adjusted to the dark, or night weakened, or all three at once, and I half saw, half felt a low barbed wire fence across my path. I pressed the wire down with my hands and swung a leg over it and, as if I hadn't enough trouble that night, I heard a heavy, invisible animal pounding and snorting inimically towards me. My leg swung smartly back and, whatever kind of beast it was, stopped and stood panting on its side of the fence while I panted with mixed fright and anger on mine.

This rather absurd episode has stayed in my memory like a darkened magic lantern slide, and click, in the next slide the night has begun to grow lighter and I am walking along a path with trees on each side. I am tired – after all I had spent a lot of the previous night bombing Hamburg, and hadn't slept much after my de-briefing at the squadron, and had prepared for, and come on the caper which had got me where I was now; and, click, another slide, I am in the grey light in a wide ditch, dry and deep with high shrubs above me, and curl up in it on my right side, my back against a grassy bank; and click again, I awake in the early morning, chilly in the ditch, unable to stop thinking about the telegram winging home, and I try again and again to send telepathic assurances that I haven't been killed.

Time passed slowly in my ditch. An hour after the sun was up I heard distant voices beyond the bank I had been facing when I slept. I crawled up it and through a gap in the bushes saw a man and woman about a hundred yards away working in a sunny field. I slid down again and waited all the day, increasingly thirsty as the sunbeams through the leaves and the shadows of the bushes crept along the bank and across the ditch and back the other side, and faded slowly, and at last the twilight came.

Click once more, and I am standing in my second night, faintly moonlit, next to a small stone trough. I suck as much as I can and fill my rubber bottle.

That night I walked for hours down empty lanes and over moonlit fields, always southwest as well as I could on my luminous compass. The stars were out, the moonlight eased the way.

Click, and I am walking down a shadowy slope of grass to halt on the straight border of a silver shining canal in front of me, and fifty yards across to the dark bank the other side.

I take off my boots and all my clothes. To lose any would be a disaster. So in the moonlight I spread my shirt on the black grass and pile on it boots, socks, underwear, battle dress, and wrap it around them, and tie the bundle firmly together with the shirt sleeves. With my right hand I press the bundle on my head for my left arm to swim side stroke across the canal, and walk into the water. It was cold around my ankles, sandy mud between my toes, my bare bright body floodlit by the moon for any foe to see. Another step and the water reaches up to my calves, and another, and I go no deeper, and another, and another, and continue my white, naked walk right across the water which never comes above my knees, and climb the bank the other side.

I dry my legs with my shirt and dress again. I climb the bank and wade through a hundred yards of bushes, black shadows, deep grass, and on a sudden break into a clearing, and a huge silver river fills the foreground, its other side far away in the moonlight.

To walk along the bank for a bridge would be folly, as it could be guarded. Exasperated, I stripped again, made another meticulous bundle and, holding it on my head again, waded into water which was immediately out of my depth. I began side stroke with my left arm, drifting downstream with the current, slowly swimming towards the other side. The river was going to be very wide to cross with one arm only. But I had swum long distances since I was a young boy, starting from beaches to boats and back – my brother and I lived at Leigh-on-Sea in Essex after our parents divorced – and later across bays, and miles along rivers at Oxford, always purposeful, to a target.

The water was not really cold. I found a relaxed rhythm despite the awkward bundle on my head and kept on, soon half in a dream. As my left arm pulled me through the water again and again, I recalled a commotion I had caused during my RAF ground training in the summer of 1940 at Blackpool, which I swam at high tide in a fairly rough sea alongside the great Blackpool Pier to the end and round and back again the other side. Two men were shouting down to me from the pier, and I only learned from them when I climbed onto the beach again that they had been trying to tell me that the waves could impale me on one of a hedge of iron stakes which I did not know had been fixed in the sand below the surface of the water in defence against hostile landing.

In my present river, I imagined again how it would have hurt if I had been spiked that way, and my underside left ribs started to feel nervous in sympathy, as if I had reason to believe there was an iron stake or a crocodile in the water beneath me, stretched out as I was in my side stroke.

Mildly disturbed by such a dozy fantasy, I went on swimming, it must have been at least half an hour, by now in a near trance. At last my feet found a muddy bank and I climbed up through reeds. It had grown darker now as the moonlight had faded. I dressed again, minutely careful to lose not a piece. Then I went plodding on across the fields until I found a country road which blessedly ran more or less south-west down which I walked for the rest of the night.

I had eaten nothing since the first evening in my dry ditch when I had sucked away the barley-sugar sweets of my escape kit. I drank the river's water from my rubber bottle and could find no more. As early night began I crawled into a thick plantation of fir trees, found a deep, comfortable bed of dried needles and fell at once into exhausted sleep.

*　*　*

The next evening, the start of my third night of this evasion, parched and profoundly hungry, I am on a country path and, click, there is a small road sign which it is too dark for me to read. I reach up and the letters are like Braille for me and the last three are 'pad'. This obviously means 'path', but the German is 'Pfad' so I could be in Holland.

I find water in another trough and fill my bottle and drink from it and hear myself – I shall never forget it – saying in my head: 'Here, hold on, save some for me,' and I realise that something will soon give if I go on in this way.

Click again, and for the last time. In the deep twilight a

small man is standing by a cottage with a light in the window. I never flew with a pistol as, on experiment, I had found it troubled the compass. But as I neared the man I thought that I could deal with him by a cross hold round his neck if he proved hostile.

What a joy when he proved to be Dutch and friendly! The German frontier could not have been far away, and he understood my German well. As soon as I said I was a British officer he pulled me into his cottage. His son was there, two or three years younger than me. They gave me bread, they gave me ham, they gave me beer. I had a big wash and shaved with my host's razor. They talked excitedly with me for hours, they were longing for liberation. They took away my dangerous battle dress jacket and gave me a brown cloth one. They gave me a packet of cigarettes which were a delight. They gave me a razor and a piece of soap to carry, they gave me sandwiches to take with me. When the early morning came, they took me to a narrow country road which ran in my direction, gave me their blessing and left without giving me what, from their point of view, might have been death to give, their identity.

No longer an unshaven fugitive RAF pilot in battle dress in enemy territory, but a tidily dressed young civilian, I felt cheerful for the first time since my plane caught fire. I walked along a tow path by a small canal on my left on which the golden water of a sunrise danced. Big fields on my right had brown and white cattle grazing on their bright green grass.

And appropriately, things kept coming my way. A man on a stool was milking one of the cows into a bucket, and he good-morninged me and I him. He asked who I was, I told him I was an English officer, and he scooped a milk can into the bucket and I drank the body-warm milk until I could drink no more.

Soon afterwards, a gamekeeper in green breeches and jacket came along the other way with a shotgun under his arm and a tall man in worn country clothes beside him. 'Good morning' they said in Dutch and I in German. The tall man asked if I was English. I told them my tale, and they talked together in Dutch. Then the tall man said the gamekeeper would help me if I went with him, and bade me good day.

The two of us walked into nearby woods and through the trees for a full half hour. Then the gamekeeper stopped by a sapling spruce some six feet high. He pulled it up, and with it a trap with dried needles glued thickly on it, showing a ladder to climb down.

He told me it was a hide which he had made for himself in case the Germans ever came after him. At present there were two Dutch boys in it who were on the run. But they would squeeze me in and I could stay there until he arranged for guides to help on my way.

It was a low room dug underground, well built with log walls, and a ventilating shaft covered in the roots of a big tree. There were two bunk beds and it was lit by candles, a scene from Peter Pan. The boys gave me a sandwich of smoked ham and a bottle of beer and we began to talk animatedly, I lounging on a bunk, and in mid-sentence I fell asleep and when I awoke they said I had slept right round the clock.

I rested underground for two more days and then another, waking fit, well and mildly irked by confinement until the gamekeeper came in the afternoon. He had arranged for guides and would come back before daybreak.

He came, I said good-bye to the boys, and we climbed the ladder out of the candlelight, up into the night and the sudden, strong scent of pine trees. By covered electric torch light he meticulously replaced the trap and its sapling, I followed closely through the ebbing of night and then the start of dawn, pushing weak light between the black tree trunks and branches of the forest. In about half an hour we stepped onto a path where two other gamekeepers were standing in early sunlight with three bicycles. We all shook hands, no one spoke, and my friend turned back to the forest.

The three of us bicycled westward through flat country-side without a stop for four hours of lanes and by-paths. Then we turned through a gap in a high hedge on our right along which we bordered a huge field of stubble. A left turn, deep into aromatic, sun-beaten farmland and we came to a large, solid-looking haystack.

We pulled our bicycles to the back of it. There, a large blonde Dutchman, with pale blue eyes, an open white shirt and a straw hat, was lolling in the sun on a deck chair in front of a straw covered door ajar to a hiding room in the hay stack. After an exchange of salutations, my two gamekeeper guides pedalled away, leaving the bicycle I had used all morning to enable me to continue my journey.

The Dutchman spoke good German and English. At first we exchanged the usual chat about when the liberation might come, and drank beer and ate a pile of sandwiches like any peacetime English picnic. I noticed that he showed none of the controlled excitement of the others who had helped on first meeting me. He told me that some other men would be coming soon to bring me to a crossing point to avoid the German frontier guards. Once I was over the frontier, he advised me to make my way to Turnhout – he gave me a sketch map – and there to go to a monastery, whose name I have now forgotten, knock on the door, state a code name – now also forgotten – and say who I was, and the monks would hide me and help me on my way.

Another half hour of chat, in which he asked me a few questions about my life and, at his request, I gave him my name and address in England. He praised the Royal Air Force, and I praised the Dutch Resistance. At twelve thirty sharp, another couple of middle-aged gamekeepers arrived on bicycles, dressed in green, like the three previous ones, with the same twisted yellow moustaches and grey to blue eyes set in shrewd wrinkled faces. They seemed like quintuplets with their fellows in my memory, determined, slightly tense, and for good reason: I was dangerous cargo.

We started at once for three more hours of lanes and woods, and finished again in green forest, sprinkled gold by the sun, exactly like the other forests of my previous week, different only in being a guarded frontier to cross into Belgium. In a dense area inside the trees we left the bicycles.

We walked forward in absolute silence, listening for a few minutes until one of the two keepers led the way into a small clearing. He pointed the direction for me to take by the sun, now shining through the trees, and whispered that when he gave me the signal I must run like hell on a straight line for five minutes.

I nodded, he looked at his wrist watch for a long, full minute, he raised his other arm, still looking to his watch for a few more seconds, and then, like any sports master at school, flashed it down. I waved to them and began to run as enjoined, supposing that they knew the routine timing of the patrols. And so it seemed: I tore between the trees, keeping the prescribed steady angle with the sun. There was no sign of German guards, no shouting or bullets round my ears, and I kept up the sprint for five minutes by my watch. At last I plonked down in the undergrowth to regain breath and start my Belgian chapter.

5

Belgium –
'I'm a British Pilot'

I walked on a compass path according to the sketch map for a couple of miles to win free from the frontier forest into open Belgian countryside, and south-west another mile or so up and down slow hills. Then came a series of lanes in the general direction of Turnhout, long lanes down between huge hedged banks for more miles, then a country road, an open hillside and the town below it.

I was soon in the town where I found a large café with a crowded terrace and bought some beer with a note from my escape kit. Then, after another scrutiny of the Dutchman's sketch map, I found my way to the stone wall of the monastery and rang the bell by a massive door of wood and hammered black iron, centuries old.

It opened, and a monk about twice my age stood in the doorway. I stated the code word, said I was a British officer and had been told they would help me. He looked left, and he looked right, and turned furiously on me, snapped that they didn't like jokes of that kind, and slammed the door.

It was the first rejection. For a second I stood numbed. Then I swivelled on my heel and walked at a fast pace, south-west as always, to get away as soon as possible, through and out of the town, my heart beating with disappointment and anger.

Just outside the town, a cross country tram going south-west stopped near me for some people to get out. I jumped on, paid for a ticket, I forget to where, and sat on one of the two wooden benches which ran facing each other inside. Through the opposite windows I watched the afternoon swing by, fields and trees behind them, segments of panorama, rows of little houses, humming to myself that every mile was one mile nearer, and on that imprecise proposition eventually calmed down.

When I later learned more of the German occupation I understood that such trams were frequently searched and, without papers as I was, that would have been the end of the run for me. But no search occurred.

Some time later a young man, roughly dressed, boarded the tram and took the only remaining seat which was opposite me. Almost at once he began looking intensely at my boots. These were indeed unusual for Belgium at this point in the war, best English black leather half boots, we called them Blüchers, made for me by Duckers at Oxford with a half inch lift on the left to offset damage to my hip in the smash up eighteen months ago. His staring made me uneasy and, as the tram indicated another town ahead, I got up and out, and made away.

Soon I was walking south-west along a tree-lined road. Slightly up a hill, round a bend in the road, and slap ahead of me a dozen or so young Germans in trim grey uniforms came walking towards me. They might have been an officer cadet course, laughing and talking together like all other students in the world. One or two glanced at me but none slowed down and my spirits revived: if things could go as smoothly as that, why not again and again until I could somehow make contact with the organized Resistance which the Intelligence officers attached to the squadrons had described so glowingly and urged us to seek out if shot down?

I saw few people in my south-west trek through side roads and lanes and each person looked as if I might do better further on. On and on, and it was near twilight, my legs like lead, that I found myself wondering whether I should have spoken to the young man who stared at my boots. Perhaps he was also a young British pilot who had been shot down, obtained civilian clothes from some well-wishers, and was in the tram hoping to get to Brussels and then home, thanks to the Resistance. Meanwhile perhaps he was wondering how I had managed to get Blücher boots of good black leather, so scarce in Belgium at this point of the war, and whether he should open a conversation with me. Perhaps he too had been to Oxford, I thought, and also had his boots made by Duckers, but left them at home.

As I shook my head clear of this waking, walking dream, there was a boy in overalls; he looked sixteen, sitting sprawled on a grassy bank by the roadside where a turning gave onto a yard and a big farmhouse.

He looked at me. I stopped and looked at him. After a moment of silence I said in French: 'I am a British pilot.' He jumped up at once and put his hand on my arm. He was pleased and excited and pulled me towards the house, looking about him. When we were round the back of it he asked me to sit on a bench for a few minutes. He said he had to go and tell 'Madame'. He knew she would help me.

Five minutes later a woman of thirty or so, also excited, pleased to see me, and frightened all at the same time. She and the boy bustled me indoors where we sat in a big farm kitchen, joining two pretty little girls, her daughters. She said her husband was a prisoner of war in Germany. She would find a way to help me. We would now have supper together and then the boy, Jean-Marie, would take me to a hut in the fields to sleep. It would be too dangerous for them to keep me in the house. I must stay inside the hut while arrangements were being made to guide me on my way. Jean-Marie would bring me food. I must eat it all immediately, and he would leave at once with any plates or other traces of my being fed there. I must promise not to leave the hut at all. If I was caught there I must promise to say I had found the hut and walked into it for a rest. She said the Germans would kill her and probably the children if they knew she had helped me.

Under those conditions I stayed in the hut that night and until late the next afternoon. Then Jean-Marie took me on a walk through fields to where he had brought two bicycles near a foot path. He said he had been to see his uncle who worked in Brussels as a butler to a prominent lady, also in the Resistance. They would help me get back to England. I was now going to be hidden in a house from where his uncle would fetch me away.

On the bicycles we worked our way for a few hours along lanes, avoiding even villages by detours which he knew, and then a country road to a wayside inn. The lady of the house, middle-aged, strongly built, with bright cheeks and blue eyes, chattered with Jean-Marie in French, looking frequently at me. He left, after multiple hand shakes, and she took me upstairs to a guest room with a massive farm bed of dark shining wood and a deep white duvet. On no account was I to leave this room, she said, which, appropriately, had chamber pots in a dark little cupboard.

She said she would bring some dinner up later on. Her husband was a prisoner of war, and she hated the Germans. I mustn't worry, her friends would help me get away. I looked too young to be a real pilot, she thought they would look more serious. I must be tired, I should get into bed and have a good rest, I should give her all my clothes and she would wash the linen and iron my coat and trousers ready for the next day. Here was a shirt to sleep in, it was her husband's, he wouldn't mind at all, he would like her to help

45

an English pilot. Did my mother like me being a pilot? I looked too young.

It was the first real bed I had climbed into for ten nights since taking off for Duisburg, by now a distant dream. The clean linen sheets stroked my skin, the pillow and duvet and their white covers rolled all over me, this warm, embracing softness carried me at once into a heavy sleep from which my hostess woke me for dinner two or three hours later. It was a generous bowl of soup, bread, roast pork and red wine. When she came to take the plates away, she said I had to turn out the light again and sleep some more. No wonder I needed sleep so much. She would bring me some coffee in the morning and, for certainty's sake, she turned out the light herself on leaving the room.

Jean-Marie's uncle came for me next day. He was short, with receding hair and an olive skin, a forgettable face and a mild demeanour. As he drove me in a small car from the inn and its winningly kind mistress, I had it in my mind that he was risking in the process, as she had too, torture and execution.

We came to Louvain and stopped at a big terrace house owned by two Belgian sisters. They were now in their fifties, and lived together in an interior, unchanged since at least the 1890s, of dark brown wood, wine coloured velvet curtains and many soft chairs with antimacassars. They were on fire for the Resistance, and disproportionately delighted to welcome my joining them at home, where I found another officer of the Royal Air Force, Squadron Leader Bastian, similarly shot down, who had also been brought by the Resistance to them. They coddled us for nearly a week while arrangements were being made behind the scenes to move us onwards.

Until now my evasion had been based on moves which I myself had initiated: the approach to the old man and his son at nightfall when I first learned that I had landed in Holland instead of Germany, and acquired clothes to travel in the daytime; the encounter with the Dutch gentleman and his gamekeeper in the sunny morning by the canal; the misfired approach to the monastery based on a mere tip; and my 'good evening' to Jean-Marie.

It was when Jean-Marie delivered me to the kind lady inn-keeper that, at first unaware, I had been assumed by the organized Resistance, and the time of initiatives was over. It was now for me to wait and obey. From the moment I entered the two sisters' house I was in the friendly but firm system which aimed to get me back to England, stage by stage.

To begin with this was a relief, a peaceful relaxation. But patience was needed: evaders like ourselves, once aboard what they called the 'Railroad' train, could not move faster than those already being shuffled down the line. If all went well, they would make the Pyrenees, Spain and the British Consulate in Madrid. So every postponement of a move, say in Paris because of a danger signal, inevitably delayed departure from Louvain. Bastian and I waited with the sisters for a long week. They were kindness incarnate, cheered us when we asked about moving on, but of course they didn't know.

At last the uncle came again one evening with his meek look. We walked to the railway station, he had already bought the tickets, and on to the train to Brussels. We stood for what seemed long hours in the corridor, he well away from Bastian and myself in case the train was searched and we were caught with no identity papers. From the railway station we followed him on foot to a corner of the Grand Place, and there into a door by a lighted shop window, up stairs to an apartment, and in to meet Georges, a dark-haired, muscular man about forty years old. After the brave uncle had taken his farewell of us, Georges lost no time in bringing us coffee and divulging that he had been a member of the Salvation Army but 'had lost the salvationist spirit.'

There were four RAF non-commissioned air crew, navigators and gunners, who had been waiting a week in Georges'
care, and we all waited a few days longer. It was a large, old-
fashioned flat. We were two to a bedroom. After the initial
excitement of being hidden in the very heart of the biggest
town in Belgium wore off – which it did pretty fast – this was
an empty period with one or two exceptions.

On Bastian's and my third day there, Georges said he
wanted to take us all to a cinema nearby. To Bastian and me
this was a grossly unnecessary risk: although we were all in
civilian clothes we had no identity papers, the sergeants
didn't know a word of French, and we would be scuppered
if we ran into a police round-up. After a strong argument he
gave in.

Georges was odd in another way too. He constantly spoke
of his Salvation Army period and his present religious
lapsing. After he found this raised little interest from us, he
told us that sometimes in the summer evenings he took a
bicycle into a park or a wood and if he found a German
soldier alone, and nobody else was about, he shot him. This
too, we did not applaud: it seemed probably untrue, mere
bragging, and, if true, a monstrous activity, considering the
terrible reprisals the Germans took at random against
Belgian civilians altogether innocent of such crimes.

Another day or two after the cinema wrangle, Georges said
a senior member of the Resistance was coming to see us. A
stocky man duly appeared. He wore large tortoiseshell spec-
tacles. When we were sitting with him in the main reception
room with his back to the door Georges, who had ushered
us together and disappeared, opened the door suddenly to
come back in. I saw our visitor snap half out of his chair and
reach towards inside his jacket, presumably for a pistol, and
relax at once when he saw it was Georges.

The gentleman said his alias was Captain Jackson, and his
message was that preparations were in hand to get us moved

to a northern part of France where the RAF landed in a clandestine airfield to bring people from London to the Resistance in Lysanders, which we knew well – aircraft with a low stalling speed and therefore able to land in small areas. Jackson said we were to be taken back to England on the return flight to relieve the pressure on the 'railroad' to the Pyrenees caused by the RAF's increased casualties and, more recently, the losses of American bombers.

He said he would soon be moving us to another safe house where we would be equipped with travel papers which the Resistance had now found a way to obtain in blank form. We would meet again soon.

And so we did. He and another man came in two cars which the Resistance had found a way of using on clandestine activity and moved all six of us across the city to a comfortable villa in the suburbs. Captain Jackson lived there with his 'lady secretary', as he put it.

On our arrival, Jackson held a meeting with us to brief us on the security measures we must observe, in particular under no circumstances to go outside. He asked us to give in any money we still had from our escape kits as it would be helpful to the Resistance who would now be looking after us. Bastian and I did this. Our four young sergeants could not, as embarrassed, they said their previous hosts had taken them out in Brussels and they had spent it all on the presents that they were carrying to give to their girlfriends when they got home.

As counterpart to this, Jackson provided plenty of cigarettes, razors, and similar necessaries, and improved our clothing as necessary. We had little to do. There were no books to read and Jackson and his secretary kept much to themselves. He was constantly going into the city for a few hours at a time, she was always at home. The atmosphere was bureaucratic. All our other helpers and hosts had been glad to see us and talkative and excited. These were not – I presumed because we were part of a never ceasing routine.

Eventually two other men, well built, well dressed, cheerful and friendly, came for us one morning. The name of one of them was Boris. They drove us to an office in the town which was in fact a Resistance centre. There we were photographed so that our travel papers could be completed. They said that soon they would be escorting us into France on a particular train on which our papers would be checked and passed by a German officer who was in fact working with the Resistance – a very brave man. Once in France, two guides from the French Resistance would take us into Paris where we would go into hiding until vacancies on the Lysanders were available.

While we were at Jackson's there were a number of visitors, all men, who stayed mostly for an hour or two in Jackson's office. Two, however, came and stayed for a few days, a young priest, with a dark-haired girl, I thought about twenty-one years old, gentle brown eyes in a finely sculptured face, graceful in movement and character, sweet and friendly towards us. They had been working in the Resistance in some other part of Belgium and had avoided capture by the Gestapo with only a minute or two to spare. They were now waiting, like us, in Jackson's establishment to be moved on their way to England.

One day they were not at the breakfast table: they had been fetched after all by the Resistance for a night convoy along the established 'railroad' for Paris and thence further. It was hard to get that girl out of my mind.

That same day, Boris came and issued us with identity papers containing our photographs and travel permits to go to Paris. He urged us to read and re-read these and memorize our names and employment details. I was a Belgian book salesman, I have forgotten my name but remember it had a particle 'de' in it.

The next morning, he and his companion came for us at

last. They had a small car this time in which they first took three of us, including me, crowded on the back seat to a café near the railway station. Boris stayed there with us and bought us coffee while the other man drove back for the remaining three. When they arrived the car was parked in the backyard of the café – Boris told me that the owner was 'OK' – and we all walked together to the railway station where Boris showed the tickets which he had already bought for us all. We went at once to our platform before the train arrived so that, when it did, we were able to get a compartment for the six of us together. Boris and the other Resistance man were in the compartment next door along the corridor, and so we left Brussels which had sheltered us.

6

Betrayed in France

The train journey through Belgium was uneventful. When we neared the French frontier, Boris came into our compartment while his friend stood in the corridor to block the view of us from any passing passengers. As arranged, Boris collected our identity papers and travel permits to get them cleared by the German officer, the collaborator, who would be there for the purpose. We were to stay seated in our compartment while this was going on and in no circumstances should we come into the corridor.

So there we sat, my heart, for one, beating furiously. Then Boris came by with a German officer who looked into the compartment and nodded and smiled encouragement at us, and I relaxed again. The train started, we rolled into France. Boris looked very pleased, it must have been a strain for him too. He came and dealt our papers back like a game of cards. He said the train would soon stop at Charleville where we were to get off as the French Resistance had sent the guide there to accompany us to Paris.

The train stopped. We followed Boris and the other man to the ticket collector at the exit gate. Boris produced our railway tickets and gave them in. We remained standing in a group on the pavement. Almost at once, a young man in a mauveish suit, who looked about eighteen, came up to Boris and spoke to him, looking at the six of us at the same time.

Boris nodded to him and then told us that this was good-bye, we should follow the boy to where we would wait for the next guides. He shook all our hands, cheerful and smiling as ever, we thanked him, and he and friend were gone.

The youth walked a few yards ahead of us. We soon turned into a broad street with high plane trees on each side, their harlequin bark bright in the sun. We were on the right side, he soon crossed to the left and we followed him to a solidly built bistro, into which he beckoned us. There was a wide entry hall where a tall man with a brown open shirt joined us at once. He had thick ginger hair on his head and bared chest. He looked rather nervous and told the boy we must go upstairs to a sitting room there.

We followed into a fair sized room with several chairs randomly placed and a small brown wooden table between two windows which gave on to the street along which we had just walked. The boy waved to us and left.

We all sat down, feeling pretty cheerful. Someone said of the train journey: 'Well, I'm bloody glad that's over,' an observation which found common assent. Then one of the young sergeants who was seated by a window said, and his voice was shrill: 'Christ, I don't like the look of this'. In one step, I joined him. Two German army trucks were halted in front of our bistro, grey uniformed troops were jumping out carrying Schmeisser machine-pistols, some running and encircling the building, others running in through the entry door. I sat down again.

Our door flew open. An officer in grey German uniform stamped in pointing a large automatic pistol at us, moving it slowly from side to side to cover us. I gazed down its barrel, it looked like a cannon's mouth, he said in a barking barrack yard voice: '*Hände hoch!*' We sat looking frozen at him, so he added in English, with a full German accent: '*Hands up!*' which was where we instantly put them. Half a dozen armed soldiers came in and stood about us.

Then a civilian arrived accompanied by the youth who had led us into this trap. The civilian wore a black top coat of some light material and strong-looking, greasy black gloves. He had short black hair and his face was bright red and wet with perspiration. He was obviously excited as he walked quickly to each of us and glared into our faces, his eyes wild. He couldn't keep his hands still. Gesticulating to the German officer, from whose uniform I could now see that he was a captain of the *Feldgendarmerie*, he said to him in uncouth German: 'I must have them for questioning'. I looked at the gloves on his moving fingers and it was clear indeed what he meant.

The captain's German had an elegant drawl, perhaps he put it on to irritate the Gestapo man, whom it was obvious he disliked. He side-stepped the request and said we must all first move to the *Feldgendarmerie*. Closely surrounded by the soldiers, we were shuffled downstairs and pushed into three or four cars with soldier drivers.

I was put on to the front seat by the driver and, to my dismay, the guard on the back seat was joined by Gestapo. His face was now writhing, I thought he was going to have a fit, and then I thought he must be acting the Gestapo role as in a crude film. At once I had cause not to give him the benefit of that doubt: an old woman stepped cautiously from the pavement curb ahead of us, the driver used his brakes, she continued walking across the street, and the Gestapo man, his face grimacing even more violently, screamed at the driver: '*Du hättest die Hündin überfahren sollen*' – 'You should have run the bitch over.'

The cars turned left from the road into a cobbled yard with a large building of grey stone on the other side. The sun was pale gold on it. At the right corner a white painted flagpole reached well above the roof, dominating us and the court-yard, with a huge Nazi flag floating from it. The bright red around a white disk within which a black, spidery Swastika seemed to be cart wheeling in the rippling wind. The message was depressing.

A sergeant with medals on his grey uniform stepped out of the entry door. On his command the guards hustled us into the building and a stone walled room with one door and large barred windows. Again, there were plenty of chairs. We were ordered and signalled to sit down and keep still.

I still had a full packet of cigarettes left from Jackson's largesse. Moving gently, I took one out of my pocket and lit it. The guards did not demur and one or two of my comrades did the same – calm seemed to be coming.

But in fact what came was the captain with a grim look on his face, and the Gestapo man, who was talking angrily to him, with the boy we had followed, standing behind.

Gestapo stopped talking when he entered the room and walked around looking at each of us. When he turned his face to mine again, his eyes were mad, his cheeks were twitching, his black gloves could have been soaked in blood.

'I have to question them, I say', he shouted again to the captain. The latter stood tall above him, said nothing for several seconds, and then calmly. 'They are my prisoners, not yours. I will not surrender them to you.' He paused. Then pointing to the youth: 'You can take him if you want,' and smiled sarcastically. The informer boy looked sheepish, but not afraid. Gestapo turned on his heel and walked out furiously, the boy behind him.

The captain rapped an order to the sergeant, the meaning of which I missed. Then he too turned and left the room, and again the quiet reigned. The sergeant sat down, his men stood about us, we stayed sitting, each, no doubt, confronting the future with uncertainty.

After an hour of this the sergeant looked at his watch, uttered a German expletive and: 'We might as well take them into the courtyard and shoot them now' he said to the troops who stood covering us. He nodded his head at a window and

55

said 'It's got to be done soon, anyway.' One of the soldiers grunted what seemed to me an approval. The others went on looking inertly at us.

My heart started to beat fast. My companions had not understood a word of what the sergeant said. I decided to leave it that way. I lit another cigarette and inhaled hard and, as my heart continued to pound, I concentrated on praying for help to die with dignity.

7

Interrogation –
The Traitors Identified

The sergeant got up and left the room. Ten minutes later he came back with his captain. The captain said he had just learned we were British flyers. We would be transferred to the *Luftwaffe* as prisoners of war. We were driven across the town that same evening to a former school containing *Luftwaffe* troops.

When the *Luftwaffe* took us over from the *Feldgendarmerie*, we were kept under guard for nearly a week in two adjoining rooms with barred windows on the third floor. Surprisingly, none of us had been searched since our arrest. We thought it best to incinerate the false identity papers we still carried, in order to forestall any dangerous ideas that we might be spies. Fortunately, we were able to do this in a small stove in one of our rooms out of sight of the guards, who allowed us to move between our rooms and a bathroom on the same landing, where they stood with rifles day and night. The *Luftwaffe* rations brought to us were adequate. We had nothing to read. One of the guards, a young corporal about our own age, lent us a pack of cards. He had brightly blonde hair and spoke tolerable English, which he told me he had learned as a paying guest in Croydon in the summer before the war.

We had speculated endlessly about who had betrayed us. The young man who led us from the station was obviously in the know, witness his lack of anxiety when the captain who had arrested us offered him as a victim to Gestapo. Presumably the innkeeper too. Were Boris and his friend arrested after we left them with the boy? If so, what about the German officer who looked at us on the train? Was he a double agent for the Gestapo?

The days wore on in desultory conversation. This included an inept idea of slicing and plaiting three or four blankets into a rope for three of us, including myself, to slide down from the bathroom window, which was not barred and gave on to a street two storeys down, while the others somehow distracted the guards. Luckily we did not proceed. Most likely a semi-alert guard would have had some nice target practice shooting down the rope.

Almost a week had gone by, when, one morning our guards walked us out of the gloom into the street. We were blinking in the sunlight, below a sapphire sky, white billowing clouds were floating overhead. I returned his playing cards to the young corporal with thanks in English and he nodded politely to me as always.

While we waited by a railway siding, I heard him say to his troops in German: 'If any of them tries to escape shoot him in the stomach,' his blue eyes merry, his pink smiling face, looking at me.

Not surprisingly, nobody did. In two adjoining compartments, the train took us through a sunny morning to the small town of Laun where we were marched up to its ancient castle. There a part was manned by trained jailers dressed as German soldiers. We were locked on the top floor, plenty of light, moderate food, and rifles everywhere.

We were told that we were going to the '*Dulag*', the *Luftwaffe* interrogation for all fallen aircrews, British and American. After that we would be shunted to our definitive camps, officers and NCOs separately.

A couple of days later, another escort took us to the 'Dulag' by express train via Frankfurt. There we slept all night on the bare concrete of the great station. The guards told us that two nights before the RAF had bombed the city, and that if we were caught by civilians it would be the end of us. In the morning we took a train for twenty minutes to Oberursel, a small town at the foot of the gentle mountains of Taunus.

*   *   *

Two months after war broke out, one of our senior officers, Wing Commander H. Day, 'Wings' Day, was shot down by German fighters, and by Christmas there were nearly twenty other captive pilots, British and French. They were held in an unpleasant country house in the Taunus. A few weeks into 1940, the *Luftwaffe* enlarged the restaurant of our past skiing holidays to become the first serious Kriegie camp. When more arrived, the inn was expanded for them, and then barracks were provided. Some officers tried to escape, but in a disorganized way. 'Wings' Day, as the Senior British Officer, quickly laid down that all escaping efforts had to be cleared on his authority. When further camps were created in different parts of Germany, the 'Wings Day system' of escape committees prevailed in them all.

The Taunus camp became the *Dulag*, the place through which every Kriegie had to pass to enable German officers to try to gather military information, before their new prisoners were sent on to their permanent camps.

*   *   *

In 1933, when I was thirteen, my father disliked my private school in Essex and sent me to stay with German friends in Frankfurt. I attended the Goethe Gymnasium for a year between schools in England. As I was the only foreigner in the whole school, I learned German fast and had many

cheerful times in the Taunus with the other boys there –
picnics or skis, and visits to a small inn in the hills. I vividly
remember breaking one of my skis (in those days they were
single wood) on trying a minor ski jump, and having to walk
down again to the train. It was the second year of Hitler's
accession to power.

* * *

On my arrival now at the *Dulag* I first remember being
totally undressed and searched in all possible ways. My civil
clothes were thrown away, the only thing I was allowed to
keep were my boots when I showed them that my left foot
needed an uplift of a half inch. I was then given a British Air
Force kit. Then a German officer was saying to me with a
stern look, 'All the power of the British Empire cannot help
you now, Mister Sniders. Was it you who bombed Duisburg,
or nearby?' I gave him my number, rank, and name, and said
'You know that is all I'm allowed to say.' He replied 'Things
would be much easier for you if you would talk a little more'.
We argued quietly for some minutes. Then, on a sudden, he
threw down two photographs on the table in front of me.
They were of a young sergeant pilot and his navigator of our
squadron whom I knew intimately. They had been missing
for some weeks. My expression became shocked. He sat back
and laughed and said: '139 Mosquito Squadron is a very good
squadron, don't you think?' and that was the end of our
conversation.

I was then put into a small cell for a day, with nothing but
cooked potatoes to eat. If I banged for a loo it was granted.
After a troubled night, I was awoken by a young German
officer. He asked who had helped me get from Duisburg to
my capture. I said I had acquired some clothes from a boy
in a field, and I did not know who he was. The officer said
he could send me to a camp where there would be very little
food. I said I really was telling the truth and for some good
reason he laughed and went away. I had expected all kinds

of bad trouble, but next day I was simply put with others into the train for the long journey to *Stalag Luft III* near the small town of Sagan, some hundred miles south from the Baltic.

When we arrived it was a sunny morning in what is now west Poland, the province of Silesia in the Third Reich, half an hour from the camp. About seventy of us, officers of the Royal Air Force, disembarked from a train of cattle wagons. We had sprawled on their floors of straw for two days. Since our various planes had taken off we had all been travelling, one way and another. Now we were walking towards camp with guards alongside us, and soon we would halt to be locked up for the rest of the war, a gloomy prospect.

It was hot along a grassy path towards the *Kommandantura** barracks. Nearing it, we walked below parasols of maple trees, their outer leaves shining bright yellow from the sunlight pouring down. There was similar greenery up to the guarded gate which was standing wide open to receive us. A huge Nazi flag hung from a flag-pole. Every hundred yards along the wooden enclosure stood a twenty-five foot high watch tower and on it a guard, a search-light and a machine gun. Through we went and the gate clanged shut.

We walked into a grubby stone yard, and then over grey sand and top soil which carpeted the compound. The only green was sparse tufts between poles which held the wire which held us. Not one petal of a flower was to be seen. Dark green and dense black conifers surrounded the entire camp which had been excavated from the forest on Goering's order a year ago, to be a model camp for captured airmen and opened in April, 1942.

Early in June, 1942, the Germans had shifted hundreds of

* *Kommandantura*. The large area for the Germans on the west side of the camp.

61

Royal Air Force NCOs* from the camp to make way for the increasing flow of shot down American officers, leaving only a small number of NCOs, who had volunteered to stay for household tasks.

By the time we got there, the camp had Centre, East and North Compounds. North Compound was invisible to us because the *Kommandantura* was between us. Ours was the large Centre Compound. The East Compound was slightly smaller and linked to us by the double wire, through which we could see and talk to our friends.

In the camp, there was a British de-briefing interview immediately on arrival. It was for me to report to the Senior British Officers' Committee how, and with whom else I had come to be shot down; to give my account between then and when I was captured for the record later on, in my case over three weeks; and to relate the questions put to me by the German interrogating officers in the *Dulag*, and my answers to them.

Another purpose of the Committee was to make a preliminary character assessment, and to place us in appropriate messes to begin with. Also, though such problems were extremely rare, they had to be sure we were what we claimed to be. Now and then in some camps the German Intelligence had tried to plant agents whose English they deemed good enough. Similar arrangements were held by the Senior American Officer and his aides, for the batches of shot down US flyers as they arrived.

My interrogator was Squadron Leader Anderson, DFC. To my immense relief, I learned from the Committee that my navigator, Squadron Leader Price, was still alive, and later in the day I got over to him. He had been captured quickly and treated unkindly by police, and then had a difficult patch

* Non-Commissioned Officers, mostly Flight Sergeants and Sergeant aircrew including pilots, navigators, wireless operators and air gunners.

through the Dulag. But he had settled now at a rather senior level, and was trying to make the best of things like the rest of us.

Anderson told me that about ten other officers had arrived in the camp in the last months who had been betrayed in the same way as we had, though the arrests were mostly made in Paris. The culprits were Jackson and his mistress. The escape line had been penetrated in Brussels by the Gestapo. Jackson's set up was controlled by it. I finished my report with my interrogation in the *Dulag*. The Chairman thanked me for my report, I stood up, saluted smartly and left their hut.

A few paces away, shock brought me to a standstill for a long moment. What Anderson had just told me about Jackson was appalling, truly terrible news. Not that Jackson had double crossed us, that was just a bit of bad luck, like a bullet in a battle. For the Gestapo we were only a minor bonus who had to be handed over to the *Luftwaffe* fairly soon. Their prime purpose was to catch members of the Resistance on the run. And Jackson was in it and methodically passing them to the Gestapo who passed them to their specialists in obtaining information.

I could not bear, yet could not stop thinking of, what by now had happened to the bodies and hearts of the lovely young girl and the brave little priest who had not been at the breakfast table at Jackson's that morning.

Grief grabbed between my lungs, my eyes were full of tears, they come again as I write this, I see her still, she is part of my life. After the war, a friend of mine who had served in the Special Operations in occupied France, organizing British air drops of arms and supplies to groups of *maquisards*, told me the story of an English girl he had known, who was captured on parachuting into France.

When France was liberated it was discovered that she had not died of her torture under interrogation but had been repeatedly raped thereafter in her Gestapo prison. Once a baby was clearly on the way, the guards left her alone for the weeks until her labour began. Then they strapped her tightly in a strait jacket and kept her so, visiting her from time to time until she died after two days and nights.

8

Escape

Two hours later, I was alone on my first afternoon in a prisoner-of-war camp for officers of the Royal Air Force, *Stalag Luft III*, Centre Compound.

A narrow length of blue-black metal tape lay on the dreary sand, shining like a snake next to the empty carton it had bound together, the start of a major adventure.

In a situation where it was evident that practically everything was scarce, I thought that that strip of metal might somehow come in handy. I picked it up and took it into the barrack hut to which I had just been told to go by the Reception Committee.

Our barrack huts had wooden walls and roofs and stood on timber piles up to a foot and a half above the earth. Inside them messes* of ten officers were divided by sacking curtains. A curtained gangway ran from one end of the hut to the other. In the centre of each hut there was a massive, tiled coal stove on which we would depend in the freezing Silesian winter. Ours was four cornered and one and a half metres high. Horizontal bits of black iron around it and vertical metal corner bands from top to bottom helped keep it together. All our water came from taps in three or four

---

* In the British Armed Forces, a group of people who generally take their meals together.

wash houses over underground pipes from the general water-mains of the camp.

There was nobody else in the hut. I wrapped my fingers in waste paper lying on the floor, grasped the metal tape with an inch of bare blade between my fists and started rubbing its edge up and down against one of the vertical metal corner bands of the stove, leaning my weight on it, rubbing hard, up and down, until I had worn a triangular indentation in the blade. Then, exactly adjacent to this, I rubbed another indentation to form a saw's tooth, and then another, and another, working for an hour or two when the hut was empty during the next few days, until I had a narrow saw blade almost two feet long.

I showed it to one of the RAF aircrew sergeants, a good carpenter, who worked for us in the camp. With a hammer and a nail he made a hole through its ends. Then, with two pieces of a broom stick, a wooden cross bar, plaited string, a twisting peg for tension at the top, and the blade fixed by nails at the bottom, he gave me a robust cross saw. I hid it under a plank under the floor beneath my bed until it would be useful. This came two months later, when it was used to make a ladder.

Inside the barrack huts, as I have mentioned, our messes were separated from each other by hanging sack cloths to provide a measure of visible, if not audible, privacy. Within each, our tiered bunks served as walls. We left space for a curtain into the central corridor, and the outer wooden wall of the building had a window for each mess, shuttered by the guards at night. We had a table and benches in the centre of our space, and an electric light from the wooden ceiling. The floor was bare boards. There was a wooden cupboard for a food store, and most of our belongings hung somehow on the walls or were packed into cardboard Red Cross boxes. The aspect did not lift one's heart.

Our domestic routine was simple. Every week a pair of

officers in turns, known then as 'the stooges', did all the housework for the mess – managing and working our rations, cleaning the floor and windows, fetching our water from the wash hut and so on. The rest of us did nothing except make our own beds and do our laundry.

After the morning *Appel** the camp would come to its customary life. The scene was one of officers walking briskly round the perimeter by the warning rail, anti-clockwise by convention; playing football in the large open stretches of the camp to the all-enshrouding wire; holding running matches, playing quoits over nets sent out by the Red Cross, boxing in home made boxing rings – all kinds of recreation, teeming about on the grey ground, constantly churned by all these boots and sporting shoes.

In the de-briefing, the Committee had been interested to hear that I was fluent in German. Next morning, Anderson took me for a walk around the perimeter, the standard for security conversations, and offered me a welcome duty. Under our British internal regulations – ultimately King's Regulations as applied to a prisoner-of-war camp by our Senior British Officer, Group Captain 'China Bull' MacDonald and his staff – we were forbidden to seek conversation with German officers and other ranks. Exceptions were a small selection of German speakers of whom I then became one, known as 'contacts'. Our role was to talk with Germans – guards, workmen, administrative staff who came into the camp – to enter as far as possible into friendly terms with them, to pick up information, to influence their morale, and obtain useful objects from them.

In 1943 the results were just satisfactory, taken together with the Red Cross parcels which at that time were being

* The parade, once every morning and once late afternoons in which officers were counted by the German Captain to be sure that none had escaped.

delivered at the rate of one parcel per prisoner per week. These came by German railway from varying delivery points, and were scrupulously delivered by the *Luftwaffe* to our commissariat. The British, American and Canadian parcels had minor differences, but all contained tins of sardines, salmon, bully beef or ham, butter, jam, sugar, dried egg powder, dried milk powder, thick packets of chocolate, dried apricots and raisins.

Those tins brought a needed element of joy to our meals. They also encouraged imaginative cooking – ways were found to mix milk powder with 'goon' margarine into what I remember as delicious whipped cream or cakes made of ground-up biscuits (the Germans banned flour for us as it was too easily used in combination with chocolate and butter to create compressed escape slabs.)

Since each mess had to do all its cooking in forty minutes on the communal stove fed by coke briquettes, in considerable discomfort, wits were sharpened to match the case and my memory is now that in 1943 all our meals were like treats.

We had plenty of cigarettes from our Red Cross parcels, and by 1943 cigarettes were scarce in Germany. We bartered them against trifles: screws, a nail brush, and, as the link developed, illicit but apparently harmless things like German cash for cigarettes, and indelible pencils. The lead of these pencils, ground up, made a dye for turning American khaki or Royal Air Force blue jackets and trousers into civilian clothes, and the money was needed when escaping, especially by train.

Among our visitors to the Compound the most frequent were three low-grade members of the camp's *Abwehr*, the Security Service, normally dressed in navy blue overalls and three-quarter length boots, whom we named 'ferrets'. They were meant to wander about the camp in and out of buildings and all over its open spaces, prowling and poking for signs related to escaping, and in particular for sand and earth lying about which looked fresh and could have been dug from a tunnel and inadequately dispersed.

In addition to our general role of suborning, now and then it was our particular function to keep them away from any area in the camp which we were instructed was sensitive. To this end we would buttonhole and talk to them as interminably as they would let us about anything under the sun, their families, the war, what did they do before the war, what would they do after it, the weather, any damn thing to keep them distracted.

Of the ferrets, as I got to know them, swarthy black-haired Peter, twice as old as us, say forty-five, proved to be heavily depressed about the war. Kurt, who was in his thirties, pink faced, fair-haired, told dirty jokes, talked about girls at home, and generally gave a slippery impression. Both of them committed mild offences in return for cigarettes.

The third would not give us his Christian name and said he was *Korporal* Schulz. He was in his fifties, a member of the Nazi Party, which the others were not. He was far from fanatic, a small shopkeeper in civil life who had probably joined the Party for safety's sake. He was quite friendly and always polite. But he refused to do any trading and often walked away from one after only a brief conversation to pursue his proper ends.

I did not think any of the three had seen action. Peter and Kurt complained about hardships at home for their families, Nazi Schulz was more reticent. None had any culture, all three could have been worse.

Soon, and unknown to the ferrets, there was an act, which would have been deemed by the German authorities to be treason, and was of high bravery. I still have two small photographs of myself in forbidden civilian clothes taken by a German technician who had work to do in the camp, was opposed to the Nazis, and agreed to bring his camera on the next visit and photograph me. On his third visit, he gave me

69

my picture and took two pictures each of half a dozen other prisoners to be used in false papers in a projected escape. He would take no reward. His discovery would have brought certain death for him and probably torture by the Gestapo first. It was his only available act of anti-Nazi resistance, a flower of courage.

The Centre Compound was a rectangle of sandy ground some 400 yards north to south by 250 east to west. The only gate was in the middle of the north fence and always guarded. The boundary for us was a low wooden rail around the perimeter beyond which a clearing 30 feet wide ran along the barbed wire fences. The sentries had orders to shoot on sight any prisoner who entered it. It was called 'the warning area'. But no warning was meant to be given.

To make the inside fence, barbed wire was interwoven criss-cross and nailed nine feet high to sturdy posts like telephone poles. This was itself surrounded by an outside second fence, at least as strong. The corridor between them was six feet wide and filled, half way up, with a thick barbed wire entanglement.

Sentry towers stood at each corner of the rectangle and along the sides where the Germans thought necessary. The towers were always manned by sentries with slung rifles and in their boxes machine-guns were set to fire along the fences. At night searchlights swept the wire, guards with rifles patrolled it.

Alongside our compound there was a smaller, similar one, the East Compound (see Figure 1). A North Compound had just been built in the camp and later a South Compound, both of which were out of our line of sight.

A month went by. I learned some of the ways of the camp, found a few acquaintances from RAF flying schools, and one from a skiing holiday before the war. New friendships developed quickly, as at school, and one in particular who is about to start a long and powerful role in this story and

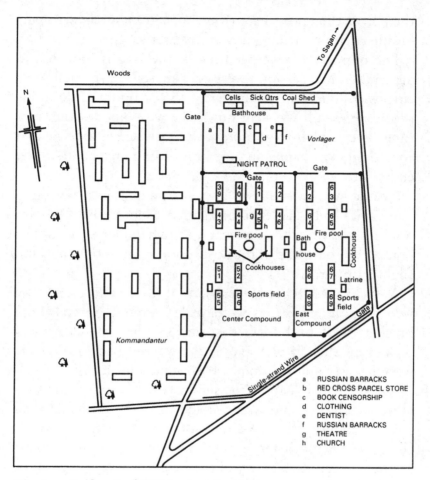

Fig 1. *Stalag Luft III* in April, 1942

whom I, ever gratefully, now salute again: Bill Manser, a fighter pilot shot down in his Hurricane on a low level strafing attack on Crete on 23 July, 1943, a week before my German affair.

Like me, Bill was too young to get into the Air Force when war began so he read modern languages at Cambridge for the first year of it. Now his German was perfect and he, too, had been doing what he could to suborn the ferrets and keep

71

them from their proper activities. Within a few pages, he puts his life at risk in helping me on my first escape.

The opportunity showed itself one bright morning in September. Manser and I were on a sharp keep-fit walk anti-clockwise round the perimeter of the compound, alongside the warning rail. We went down the west side for the third time, wheeled left to go along the south fence, and saw that something was happening in the warning area at the other end of it.

Nearer, we saw that two goon carpenters in mauve-grey work suits had a ladder tilted against a fence post. One was holding it steady, the other was up and hammering a wooden bracket on to the top of the post, pointing into the camp and slightly upwards.

Any changes to the fences were of interest to us, and we watched, on and off, as we dawdled around. Several more brackets were up by half past eleven when the carpenters picked up their ladder and tools, walked back across the warning area and the camp, we presumed for lunch.

The change of guard in the sentry boxes always took place at noon. The carpenters came back around one o'clock, and hammered away for the afternoon by which time about a third of the south fence posts had brackets on them. It seemed evident that when all the brackets were on the four fences surrounding us barbed wire would be fixed along them to make an overhang towards us.

The next day they added brackets to some twenty more posts. They took the same lunch break, and I formed an escape plan which I put to Manser who agreed to join it. It was predicated on my belief that a sentry would never open fire without warning on a man in German uniform. Under our strict regulations, I submitted the plan to Anderson and asked for clearance from the Escape Committee. He said he liked the idea and would let me know soon.

A night or two later the camp's acting enthusiasts were to

72

perform a play in the hut set aside for entertainment and lectures. It had taken months to prepare. According to a civilized practice agreed by the German *Kommandant* of the camp, Colonel Friedrich-Wilhelm von Lindeiner, and the Senior British Officer, small parties of officers were allowed to be escorted under parole from the East and North Compounds to join for the performance. Traditionally, von Lindeiner and some of his staff sat in the front row with our senior officers, as our guests.

Needless to say, the usual brief news summary from England, which came from a clandestine receiver, known as 'the Squeaker' and was managed and deeply hidden by the Committee, was not read on such evenings.

In the interval of the play Anderson sought me out. He said I was to have a talk after the play with one of the North Compound party, Squadron Leader Roger Bushell, who was the head of their Escape Committee, as he would like a second opinion on my proposal from Bushell before deciding.

I was too new to the camp to know anything of Bushell at that time, but learned much about him soon afterwards. He was old enough to have been to Oxford before the war, where he flew as a pilot in the University Air Squadron, and to the Sorbonne. He was a British ski-champion and an increasingly successful barrister when the war broke out. He was shot down early in 1941 and was by now famous for two escapes, in the second of which he was at large in Germany for six months, including some time with Czech partisans, before recapture and return to Sagan after a grim passage with the Gestapo.

This, my only meeting with Bushell, has been unforgettable. He was later the leader of the great tunnel in the North Compound through which 73 officers escaped at night in March 1944, as described in Chapter 17.

When the applause in the theatre had stopped, Bushell

took me aside and asked me to tell him my proposal. In sum, it was for Manser and me to climb over the wire disguised as carpenters after they had gone to lunch and the guard had been changed in the sentry towers. He questioned me in detail, said he thought it made sense and would talk with Anderson about it, and bade me good evening.

The next day, Anderson took me for another walk around the perimeter. He said there was one condition which the Escape Committee insisted on, namely that it could only be a one man escape. It would be essential for Manser to help as there were two goons working together on the fence who had to be impersonated. But the Committee thought that for two of us, even disguised as goons, to try to get over the wire would certainly arouse suspicion to the sentry in the box. Manser's role must be limited to carrying the ladder to the fence and back into the camp after I had climbed it. Subject to that, and if Manser agreed, we could go ahead, and we would get full support from the camp's escape organization.

When all this was put to Manser he agreed. This decision was an act of pure honour. I was the one who stood to win free if the thing worked; and if it didn't, he risked getting shot along with me.

The support we got from the escape organization was full indeed. I disinterred my cross-cut saw with the blue-black blade, and with it a ladder was made like the German carpenters' with wood purloined from the huts, and hidden in the roof. Autumn was not yet fully on us so I could travel lightly dressed. A Royal Air Force sergeant pilot, a tailor in civil life, recut and sewed an American officer's khaki uniform to fit me, re-equipped with civilian buttons and dyed a deep purple from the indelible pencils. This was to be worn underneath the copies of the carpenters' uniforms which were to be the top disguise for the two of us, and made by mild dying of RAF trousers and two good English pyjama jackets. Even I discovered an unknown talent for embroidery and changed two RAF forage caps into highly convincing German privates' caps of much the same colour by embroidering on them replicas of *Luftwaffe* winged cap badges with handsome little swastikas at their centres.

Identity papers were made for me as a young French worker, food concentrations were prepared, German money, a map painfully copied from a precious original, a home-made compass – all these and other necessities – were made, or already in the Escape Committee's hidden stores and were to be packed around me under my carpenter's uniform when the morning came and our final briefing from Anderson was done.

We decided that our point of attack should be the west fence where it was opposite our huts in the northern quarter of the compound. On the other side of the fence, thirty yards from the wire, stood the first of a few rows of barrack huts used as the administrative offices of the *Kommandantura*, surrounded by a rough lawn with trees and bushes among them. At the north end of this section of fence there was a sentry tower before the last tower up in the corner, and the way our huts had been placed blocked the view of the sentry at the camp gate.

The carpenters began fixing the barbed wire with staples to form the overhang. They carried a large ball of wire to the foot of the post in the southeast corner. One of them unrolled enough slack for the other to climb the ladder and hammer on the wire three inches from the bracket's joint on the top of the post. Then down, and the same to the next post, and so on until they took their lunch break, leaving the ball of wire.

We waited until the first strand of wire was all around the camp and the second strand hanging along the south fence and moving up the west side. The time had come.

It was a sunny day. We breathed a first hint of autumn in the air. Around eleven thirty the carpenters picked up their ladder and tools and walked across the warning area, between our huts to the gate, and away.

At noon the guard was changed in the sentry towers. The new shift climbed up, took their positions on the platforms and stood, rifles slung, looking over their parapets along the wire and over the warning area.

By now the ball of wire had been replaced many times and the one presently lying on the ground was about the size of a football. They had left it below the pole next along the fence from where they had finished hammering. This was about thirty-five feet from the sentry tower towards which they were working. We considered that was near enough to hear speech going on but not near enough to hear it clearly unless voices were raised.

Twenty minutes later, when the new guards had had time to settle in, the first carpenter in his dreary mauve fatigue uniform (myself) walked from behind a hut carrying the front end of the ladder (courtesy of the strip of blue-black metal found on my first day in the camp), and followed by another goon (Manser) dressed the same way, holding the other end of the ladder.

We strolled with maximum unconcern towards the warning area. In the lead, I reached the low railing and stepped over it without slowing my pace. Manser likewise. We continued towards the fence without looking towards the sentry in the tower, and stopped and placed the ladder against the pole where the ball of wire lay waiting.

We had heard not a sound from the sentry. Manser picked up the ball of wire and unwound a few yards of it. I got on the ladder with the ball of wire in one hand, and climbed until I was standing high enough to start work. As already described, I had convinced myself that the sentry would not put a bullet into a German uniform without a challenge. But my ears were open all the time for a shout which did not come. I pulled some staples from a pocket, the hammer from my belt, and dawdled away a few minutes hammering them through the twist of the wire into the projecting bracket.

When the third staple was nearly in I flicked a glance at the sentry box and back again. The guard was looking at the scenery. So I gave a final bang with the hammer and, with Manser steadying the ladder, climbed down onto the sand. We drearily picked up each end of the ladder, one of us took the ball of wire, unrolling as we went, and moved on to the next pole four yards nearer the sentry tower. The guard was still looking away with evidently no interest in us. We placed the ladder against the pole, I climbed it again, lifting the ball of wire with me, and hammered in the first staple to fix the wire.

By now we had been several minutes at this game, and it was time for the next move if we were not to risk being over-taken by the return of the real carpenters before the whole play had been acted out, with me away and Manser safe back in the camp.

The guard was still not looking at us. I tossed the ball across the top of both fences, and it landed in the grass a yard outside. After a pause full of listening Manser shouted at me in his excellent German, loud enough for the sentry to hear, that I was an idiot, now I would have to climb over to throw the wire back and walk round to the gate. I snarled at him and stepped up the ladder to the top of the fence, facing away from the sentry to avoid eye contact.

I swung one foot over the criss-cross of wire next to the pole, right round and into the wire so that my boot pointed back to the camp, and then the other, over and into the criss-cross a step below. Still quiet, I climbed unhurriedly down inside the first wire, cutting my hands as I went, and trod on to the thick wire tangle. I marked time on it to free my legs, treading slowly as if this was a normal job of work, painstaking as the real carpenter would have done if he had been doing it at all. My ears were ever open for a challenge, my hands ready to go up smartly if it came, but the guard remained uninterested in what we were doing. Then, with a final stamp and a tear in the calf of a trouser leg I started spread-eagle up the criss-cross of the outer fence.

At this point, concerted to distract, Squadron Leader Anderson, in his RAF uniform, walked from behind the huts up to the warning rail. He called to Manser in loud Englishman's German to bring the ladder into the camp for a minute please, to get something off a roof. The guard was now looking at Anderson with a mild interest in what he wanted. Manser was moving quietly back with the ladder as requested, and I reached the top of the outer fence, clambered laboriously over it and started down the other side.

Just before Manser and his ladder disappeared round the hut behind Anderson, I stepped off the wire into the grass outside the fence. I turned round to find no Manser and no ladder. I mimed an imprecation, shaking my fists in the air, picked up the ball of wire, slung it back over the double fence into the warning area and turned and walked away towards the huts of the *Kommandantura*. Despite my conviction, I felt an area of sensitivity in the middle of my back until there were a couple of trees and the corner of a hut between the sentry tower and me.

The whole manoeuvre, from tossing the ball of wire outside the double fence to starting through the *Kommandantura* must have taken no more than ten unhurried minutes. It had seemed much longer.

I now had to walk for two or three hundred yards through more rows of German huts, and a garden, encountering occasional German troops. Thereafter, I had been briefed, a simple garden fence marked the camp's boundary alongside the main road to Sagan, half an hour's walk away. Across the road forest began and, a mile inside it, the railway ran from Sagan north to the Baltic port of Stettin.

I set off. Soon two goons came walking towards me. As we

crossed I avoided their eyes in case they thought of speaking to me. They took no notice. It was encouraging that my uniform also passed muster at close quarters. I planned to hide it in the forest and, as a civilian, then to work my way through the trees towards the railway line, where, as other escapers had done before, I would jump a freight train to Stettin. The idea was then to get to the docks and the pubs where the Swedish sailors who plied the Stockholm to Stettin run hit it up. We had heard that they were friendly. Indeed, Anderson had said I could promise a likely Swede a thousand pounds – in those days still a lot of money for a sailor – to smuggle me aboard and get me to the British Consulate in Stockholm. With luck I could be home in a week.

9

Caught in the Act

But luck then failed me. I was walking parallel to one of the huts when who but ferret Schulz came swinging briskly round its corner, staring straight ahead and thus at once into my face before he even saw my uniform, the very face that had talked and talked to him whenever he permitted it.

'Mr Sniders, what are you doing here?', and then taking in the uniform: 'Oh, I see,' and put his hand on my sleeve.

It was maddening. One of the three men who really knew me amongst the multitude of camp staff who did not. I still feel heartbreak whenever I think of it.

He had his pistol on his belt, and there was nothing I could do, not even run for it, still deep in the *Kommandantura* and with other goons walking about. I said urgently: 'Leave me alone and you'll get a thousand pounds at the end of the war. Just walk away, and I promise you'll get it from my government.' But he looked shocked and said, stammering: 'People have seen us, I can't' which surprised me, coming from correct little Nazi Schulz.

So that was it. He called another goon over and sent him to bring help. I stood waiting, feeling a slight nausea, and a disappointment so strong that it hurt my lungs.

When the help Schulz had called for arrived, it was no less than Colonel von Lindeiner himself with an NCO. He looked shaken: successful escapes called down criticism on him and he, a gentleman of good family and an old school army officer, was in any case mistrusted in the Nazi hierarchy. He spoke at once in English: 'How have you escaped?' I did not answer. 'Is there a tunnel?' He was clearly agitated by the idea.

I said: 'You know it is our duty as officers to try to escape, Sir' which earned me a baleful look, and rightly so, for its pomposity and irrelevance. I had thought I ought to say something and couldn't think of anything better.

He saw the blood on my hands. 'Why are your fingers bleeding?' Silence. 'Come along with us.' We walked back to the edge of the compound and up along the west fence I had just climbed, heading to the Vorlager where there were guard rooms, the prisoners' hospital, and the cell block.

I saw one of our sergeants walking by the warning rail and shouted, 'I've been caught' for him to get the word back without delay for the Escape Committee to prepare for searches. The Colonel said angrily in German: 'You'll be punished for that,' but in fact that was the last I heard of it.

The cell block reached, the Colonel and I sat down in a small waiting room while an NCO and another goon stood about. He had calmed down. I was feeling full of anti-climax. A small metal stove was burning with its lid open, so I slowly worked my hand to where my false papers were in a pocket of the civilian suit under my carpenter's uniform, and snapped them out and into the fire before anyone could stop me. This brought the place to life again. Lindeiner jumped up and angrily repeated that I would be punished for that.

At that very moment a clerk came in and said: '*Herr Oberst*,* a sentry's just rung up about a carpenter who climbed over the wire a while ago and hasn't come back.'

Thunderclap and change of atmosphere. The Colonel drooped his shoulders expressively, and: 'Oh my God, our

* Colonel.

81

sentries!' was his comment to the NCO. He sat down and looked appraisingly at me for several seconds. Then, the answer to his anxiety about the tunnel evidently found, he turned to the NCO and told him to take me away.

I stood up and saluted automatically, and realized when it was too late that I was saluting with a bogus German uniform cap and a home-made swastika still on my head. The Colonel remained seated and simply nodded an acknowledgement, a little ironically but not unkindly.

10

'. . . Down Thou Climbing Sorrow'

My excitements were now over. The NCO led me down a corridor weakly lit from ceiling bulbs. There were wooden cell doors on each side with small slides at eye level so the guards could see what was going on. Except for these doors, the entire edifice seemed to be made of brick which had darkened with time and no one had troubled to paint. The gloom of all this well matched my own.

Inside, my cell was four and a half paces long and three wide. A low wooden bed was fixed to the wall, with a mattress of hemp filled with straw, two German army blankets, and a pillow with a white cotton pillow case. A wooden shelf on the wall and a stool provided sparse but reasonable comfort. The walls were covered with worn grey plaster and decorated here and there by previous guests whose small crosses were scratched in neat rows to mark the days of their confinements.

Some time ticked by while I could think of nothing but the disaster which had happened. The idea I was still in love with, my marvellous invention. It had really worked, the climb was over and I was away. Then this one and only wretch who knew me and was my polite enemy – for sure, each of my other two ferrets might have looked away if they had seen me – and no other goon of the hundreds in camp knew my face at all, this jinx, this unclean spirit, comes slap around the corner and bumps his bloody eyes into mine. If only I'd been looking the other way, if only this, if only that, damn, damn, damn, failure's litany verging on self-pity welled up in me for a miserable minute. Then I took a deliberate pull – '*hysterico passio*, down thou climbing sorrow' – and that was just as well because a guard at that moment opened the door and a trimly dressed *Luftwaffe* officer came in and sat down on a chair brought in for the purpose.

He bore no flying insignia or decorations, so I supposed he was an intelligence officer. He opened the conversation, relaxed and half-smiling, in clear, smooth English with: 'Good afternoon Mr Sniders, who for God's sake suggested you should do such a dangerous thing?'

I said, nobody, it was my idea, and he offered me a cigarette which, at that moment, I was happy to take. He tried to see if he could get anything from me about our escape organization with: 'You couldn't have done it all alone, are you sure no one pushed you into it?', and so on. I kept maintaining that it was all my own show, and when his line of endeavour was clearly getting him nowhere he stopped talking and sat and looked into my eyes in a solemn way for several seconds.

Then he said slowly as if deliberating something important: 'You must want to get home very much to take such a risk. Any other sentry but that fool would have shot you, you really must want to get home a lot,' all in a sympathetic 'my poor, dear friend' sort of tone. I did not respond to this. He said nothing for a full half a minute. Then in a serious voice: 'I think I could find a way to get you home.'

The little room turned cold and still, my heart began to beat faster, the air felt electric with treachery.

He waited for me to ask what he meant, and I did not.

Then he said, 'But no, I don't suppose you would do it' and paused again, inviting. Again I stayed quiet.

Then he said he might pay me another visit soon, and changed the subject and – surprising me again – his tone. This time it was one of approbation, saying that I 'had made a very sporting escape', I had done no damage to German property, I would have to stay in the cooler for two weeks, but instead of bread and water, my full Red Cross rations would be brought in from the camp along with cigarettes and books, and my RAF uniform from my mess. He offered me another cigarette which I took, once again gladly, wished me good afternoon, and I never saw him again.

After he left, I lay on the bed and straight away fell asleep, dressed, with the electric light on, and dreamed about nothing. A guard woke me up at the end of the afternoon. He politely handed me a sack full of the commodities my visitor had said I would get, and a plate high with hot potato salad and left me to my meal for which, as I smelt it, I suddenly felt starving. It had been a long day.

Half an hour later, the guard returned with my own uniform, and waited while I changed out of my false clothes which he took away.

The window, which faced the door and started up six feet from the ground, was covered by horizontal slats tilted so that a prisoner could not look outside. I switched off the electric light to see how much daylight came in, and it was barely enough to read by. By climbing onto the stool I could glimpse up between two slats at the bottom of the window and black metal bars on the other side. Knowing that I would soon be out and back in the camp, I dismissed any idea of seeking a way to escape from the cell block. There would soon be time enough again.

Meanwhile, my recent effort provoked an unexpected reaction from the senior guard. The morning after my incarceration I was doing push-ups on the floor when he walked in and said: 'Ah, Mr Sniders, keeping fit to try again, yes?' 'Yes,' said I, 'you're dead right.' He asked me to sit down, he wanted to talk to me, so I took my stool. He then said (and I quote from vivid memory): 'You had no right to take a risk like that. If I'd been there on sentry go, your young life would be under the ground this morning. Think of your family and wait for the end of the war. I mean this well Mr Sniders.' He was twice my age or more, obviously sincere and wishing me well. I was touched, I had nothing to say.

Indeed during this confinement the Germans from the Colonel downwards were, on balance, agreeable to me. My stunt of climbing over the wire in broad daylight seemed to have caused a sort of honourable scandal rather than indignation from the *Luftwaffe*.

✳　　✳　　✳

On my first day in the cooler I learned what I could about my environment beyond the cell itself. In the corridor there was a lavatory in a closet with a door which shut but did not lock. It had electric light, a blocked window and a cistern and chain. On my request, and if the lavatory was free, the guard on duty would unlock my door and escort me there. If it was not free I was meant to wait in the cell until the other prisoner was back in his so that we could not talk to each other in passing. But the guards were slack, and the two of us – it seemed we were the only two imprisoned in the block at that time – had a brief exchange of chatter in the corridor in the afternoon; and that happened several times during my spell in the cooler.

He was an Australian Flight Lieutenant, a tall pilot from the North Compound which the cooler served as well as our own Centre Compound. He was in worse position than me. He had been sentenced to one month, with bread and water only, for losing his temper with one of the ferrets who offended him during a search, and then blowing up the superior ferret who was called on to the scene. He was amazingly cheerful in the circumstances.

We somehow found a way for me to stuff cigarettes and some items of food at the back of the cistern in the lavatory so that he would fetch them soon afterwards. This was some comfort to him. Our present guards either did not know about or, I think more likely, closed an eye to it, as we were lucky in their particular characters.

I must have known more about this officer at the time, including his name, but my clandestine help to him, and his bright, pale brown eyes smiling at me when we met in the corridor, are all that now remains of him in the film in my head.

Indeed there remains little in that film which relates specifically to my jail sentence. After the emotions which followed my recapture had passed – if only my eyes had been looking the other way when Schulz came round the corner, and so on – I settled down and read a book or two brought over to me from the camp.

I embarked on what was intended to be an epic poem in iambic pentameters about Orpheus and Euridice, and continued to work sporadically on it after I was out of the cooler. The manuscript survived all the things which made for the contrary in the next two years, by which time it was about three hundred lines long. After the war, back at Magdalen in the Michaelmas term of 1945, C.S. Lewis was kindly encouraging, and there, alas, the matter has rested.

Further to divert myself in the cell, I formed a habit of singing loudly the old Western ballad of Frankie and Johnnie, learnt from a Rhodes Scholar, Charlie Collingwood, from New College, who used to sing it on frequent request when we had evening parties in our rooms. Charlie's good, natural baritone and hilariously heart-broken rendering was so far beyond my capacities that the guards eventually begged me to desist, which I graciously did.

There were other trivia. Although I did not think so at the time the enforced rest was probably quite good for me: it was some time by now since Price and I had taken off to bomb Duisburg, and life since then had not been without incident despite some massive slabs of tedium. I did a full pattern of exercises every morning in my cell, and generally calmed down. For a while my isolation was also a relief from living in the camp at close quarters, even with my friends.

In the passage, beyond the lavatory, there was a small window at chest level criss-crossed by black iron bars. This gave onto a low, dirty white barrack, twenty yards away and scattered with patchy grass. It was the only place in our situation which allowed one's eyes to focus beyond the rather claustrophobic cells. And I tended to stay and look at it until our guard came knocking.

One afternoon I saw a Russian prisoner plodding near and he saw me behind the bars. He looked unhappy and doubtless was, but offered a greeting in Russian which I returned in English. I threw a packet of cigarettes through the bars in his direction to cheer him up for a moment. His face indeed lit up, and he jumped and caught it, and beamed his pleasure.

Two seconds later, a goon guard, complete with rifle, yelled him to stop, strode up, and snatched the cigarettes away, shouting in German: 'You get cigarettes when I have none, you pig.' Tyranny floated on the mean air.

I knew there was a contingent of Russians alongside a part of our compound, nominally prisoners-of-war, in fact mere slaves to perform all the dirtiest jobs for the goons. Until that interlude I had not given them much thought, wrapped as I was in preparation for my own escape.

In the *Times Atlas of the Second World War* (page 205) it is estimated that of some five million prisoners taken by the Germans from the Red Army four million died, at least three million in German hands, and another million are believed to have died in labour camps on their return to the Soviet Union after 1945.

*   *   *

It is now my seventh night in the cell. I cannot sleep. In the morning I will scratch a neat little cross in the plaster on the wall next to the neat row of six I have scratched there already. The morning after will begin the second week with an eighth cross, below the first, starting a new row.

The walls of my cell weighed down. So I tried to relieve my present world, by episode, starting as far back as I could and my first syllable of recorded time was to look up from my gas lit cradle to my gigantic grandmother, all in black, looking down at me and turning to laugh with my gigantic nurse in her wine red uniform.

My first memory of standing upright was to grab the black leg of her enormous shining piano, and look up and past her playing to a crystal pear hanging on a chandelier, and the dazzle and joy of moving my head to turn the crystal to blue, green, yellow and red and back with a swivel to yellow, green and blue, and back again to green and yellow, and so on again and again until my grandmother stopped playing to look down at me and laugh again from her black and shining dress. Years later, I learned that all her clothes were black since the day that a telegram had said that her eldest son had been killed in the landing at Gallipoli in 1915.

In the next picture I was standing on tip toe, my full height in the sun, and a mass of cool blue flowers in the garden of her house in London. The last was standing straight to see her lying in her enormous Victorian bed, only her bound face and head in white sheets, because poor Grandma had died from pneumonia.

My back on the mattress, my feet on the floor, it was not the dreary ceiling but a closed vision of seven years before I was born. I saw my father in the First World War, a young man in the 'Buffs',* the trenches in the daytime, the incessant rifle fire, the deep mud with pieces of corpses in it, and the

* The Royal East Kent Regiment

night patrols and the bayonets he killed with, a long time before the shrapnel hit him which sent him home.

My head drove all day like a slow black and white cinema and would have been endlessly tedious for an onlooker. I at six, a brother and two sisters even younger. While our parents divorced, the four of us were happy in a Montessori boarding school for over three years, the Madonna at Letchworth, after which we emerged at least one year ahead of most other schools.

As previously told to our debriefing Committee, I had been sent to Frankfurt by my father at thirteen. Then I went to the City of London school spending my holidays in England, sailing with older boys on the Norfolk Broads and then rock climbing in North Wales. From about fifteen on, my holidays were nearly all abroad: visits in Germany again, plenty of skiing in Switzerland and Norway, sailing in the North Sea, visits to Paris and, for a treat, once flying there and back.

A short holiday in Germany, this time Düsseldorf at sixteen. I was sitting in the train opposite a sturdy young man with piercing blue eyes. When I mentioned Franz Marc he almost shouted, 'We do not have blue horses in the German Reich paintings'. The first echo for me of the Nazi spirit.

In early May, 1938, I had a day from school, stepping down from an early London train into Oxford and beautiful Magdalen which I was seeing for the first time. The target was to try for a Doncaster scholarship and the principal subject was French. When the writing was done, the few candidates were called, one by one, into a small sombre room. In my turn three amiable dons asked me to sit down. They said they were sorry that I had missed it this time, but I was not quite eighteen and they would like to see me again next year. Next year came and I got it. As a reward, my father sent me on the *France* to America, where I had numerous enjoyments with American boys and, for the first time, girls.

As I began to think about them, the guard brought in my dinner and the American adventure stopped. The German dinner gone, three happy terms at Magdalen came: C.S. Lewis in English literature and the stark beauty of Anglo-Saxon, a number of agreeable clubs, punting with girls on the rivers or wrestling with the swans, getting stronger in the University jiu-jitsu team; parties galore. Oxford summer term, 1940, black bow ties for our examinations, white bow ties for our dances.

I swung to the end of childhood, as I thought, until some god told me to organize a party at Magdalen in favour of the Duke of Clarence whose king and brother, Richard the Third, 'hastely dround him in a Butt of Malmsey' in 1593, now called Madeira. A few kindred spirits and I went to Bond, our Steward, superb in his craft and coming on fifty. When I explained the evident necessity it fell on fertile ears. Madeira almost never dies he said, and 'for you young gentlemen' (our average age was, after all, nineteen), the cellar would sell us a dozen bottles of 1848 at the purchase price of a shilling a bottle. As we were going to be twenty-four round the table, including a few from Christ Church and others, he found a small carved banqueting room for us, centuries old. Moreover we could use some precious college silver, albeit only of the second half of the seventeenth century, as King Charles the First had melted all the college's Elizabethan and medieval plate for money when he came through the college in the war with the Roundheads, some three and a half centuries ago.

Lying now on my cell plank and fed by Kriegie food, I looked wistfully back at the lobsters sent from London for the party, the crispest ducks from Aylesbury, the exquisite mauve-red summer pudding.

The poor Duke was toasted again and again. Malmsey was the cry until the pale pink joy was done and we were all mellow and turned to port, and livelier we got.

Then Charly Collingwood, our favourite Rhodes scholar, bellowed into 'Frankie and Johnny were lovers . . .' and everyone joined in when it came to 'There ain't no good in men . . .' This having wandered somewhat from the Duke of Clarence, one noisy spirit sang loudly to the tune of our national anthem.'

'O what a dirty dog,
Richard the Turd . . .'

We all called down on him, no more sing song, and the party chatted cheerfully with adequate port until our visitors had to leave for the minutes needed to walk back to their colleges by midnight, or be gated, or climb walls ten feet high with spikes on top, which was not unusual when returning from a party in London.

By autumn, we were all wearing diverse uniforms. So faded the things our fathers knew and our own childhoods. A few of us came back five years later, not quite the same.

11

Tunnelling

The morning two guards escorted me back into the Centre Compound I was called for a talk with the Committee to report what had happened after I vanished over the top of the wire two weeks ago.

The thing that seemed to interest them most was the fly cast by the German intelligence officer. They agreed that I had done best to leave it alone.

Afterwards, Anderson took me for a walk around the perimeter to tell me what, as a new boy, I had not known before, that a tunnel in our Compound was well advanced. Would I like to join the effort and have a place in the exit queue? I said I certainly would.

There was a lot of activity in the tunnel around the entry shaft, and it was indispensable to keep roving Germans, and especially the ferrets, away as much as possible. The Committee had established a good network of lookouts and signallers at fixed points with interlocking fields of vision and disguised signals, always manned during the tunnelling periods. My part in this, along with two similarly engaged, was to get into conversation with any goon moving in the wrong direction, and to slow him down enough for the signallers to warn those at the shaft to hide the signs of the operation according to the well devised and practised routines.

While I had been in the cooler there had been comings and goings among the messes, and the mess I was allotted to when I came back included five officers who were working on the tunnel.

Two were Canadians: Bert Larson with a huge face and big black beard, noisy with laughter, full of good spirits and very strong, had fixed a bar to our mess entry and was constantly doing sets of sixty pull-ups; and 'Mo' Macay, more reserved, a steady character. Later in our captivity, Mo received a dark blue pullover in a Red Cross parcel, well knitted by a girl he did not know. When he sent her a letter of thanks, she answered that she was very sorry he had got it, she had intended it for someone who was 'fighting like a man, not someone who was skulking behind barbed wire'. Pinned to our notice board this evoked varied comment, mostly laughter.

We had two Americans: Billy from New York, a calm, good humoured young navigator, pink face, black hair, tough inside, whom I was to have a lot to do with, one of the expert diggers at the face of the tunnel, and Jim Hardie, a red-headed Texan volunteer who got into the RAF before America entered the war. He had recently married, though barely twenty, was talkative about it if opportunity arose, and was constantly raising it.

The English contingent, in addition to myself, consisted of young Larry, also a digger, not only English but resolutely cockney, who put on being rough and tough, even used rhyming slang which was unusual for an officer of the Royal Air Force. 'What's a "titfer" Larry?' I once asked when he used the word. 'Tit fer tat, 'at, don't you bleeding know that?' – was the answer. But he was in fact a sensitive and kind hearted chap.

The tunnel had started north of one of our two most southern barrack huts. Its face was by now nearing the south warning zone. There were some hundred feet still to go:

under the zone, the double wire fence and the twenty feet clearing between it and the edge of the thick spruce forest. Three more weeks should about do it.

The entry to the tunnel was a trap disguised at the bottom of a tiled stove in the centre of the barrack hut. Lifted, the trap exposed the mouth of a main shaft through the wooden floor of the hut. That, like all the others, stood on short, sturdy piles a foot and a half high. A ladder down the shaft gave into a shaft chamber twelve feet below, dug out of the sand and revetted with planks taken from different parts of different huts.

From the shaft chamber the diggers began and, over the succeeding months, had pursued their goal of a tunnel-face well over two hundred feet away, along which they crawled to and fro each shift, passing through revetments consisting of bed boards. Such boards were normally laid across the tiered bunks of officer prisoners. Under the authority of the Senior British Officer, one board per officer had to be given up as the need arose for the tunnellers. It was an unpopular tax obtained by levy by the Escape Committee.

From the tunnel face, wooden trolleys full of new yellow sand were dragged back to the shaft chamber. It was put into narrow sacks which were lifted to the surface where officers hung them inside their trousers. When the going was clear they sauntered away to the playing fields, or other suitable corners, to dribble the bright yellow sand onto the universal gray sand and mix it beyond discovery by the ferrets. The lamps to light the tunnel, small tins of boiled margarine with wicks made of string were stored. All tools were left. And to be out of the way of all this coming and going, opposite the mouth of the tunnel a short passage gave from the chamber into a pumping room.

There, two officers sat and pumped and pumped all through working time, forcing fresh air through a pipe along the tunnel to the diggers.

All tunnel technology depended on air pipes, except for really short tunnels. The space was just high and wide enough for a man to crawl along, except for a small chamber every thirty yards in which to turn around to go the other way. Prolonged work in a narrow tunnel much more than some thirty yards long was only possible if enough fresh air could be pumped along it or down a shaft from the surface.

The Canadian Red Cross was the blessing which, all unwittingly, provided perfect air pipes for use in tunnels, with the long, slender butter tins in their food parcels, cut top and bottom, and stuck end to end.

On the other side of the pumping room there was the dark mouth of another and much tighter tunnel (the air shaft) cut in the sand two feet from the floor like an exit for a dog. It climbed back the opposite way from the main shaft to a trap disguised in the ground below the hut. Some way up the pipe, clean air was sucked down for the pump.

And so the next three weeks were spent. By common consent in the mess, the diggers, Larry and Billy, were spared housework. It was to the wash hut that they repaired at the end of the digging shifts for showers to remove all sand from their bodies. In the morning, down the ladder, they changed out of uniform on arrival at the shaft chamber into thin shirts and pants to dig in – it was never cold in a prisoner-of-war's tunnel, and the nearer its face the hotter it became.

After their work, they took a shower when they got to the wash room. The showers were icy and, by the time their jobs were done, those two well needed, and were accorded, a rest in the mess, and now and then an extra slice of bread, spread with jam or cheese, with their dinner.

That was reward indeed: from the beginning of capture every one of us felt an edge of hunger, increasing as the war went on. The *Luftwaffe* observed the Geneva Convention on Prisoners-of-War although always on the short side, food roughly on the lines of their own troops, delivered to our own commissariat, managed by British NCOs and carefully allotted to the messes.

Parallel with our meals and the progress underground, our clandestine escape factory's specialists were working flat out to make the usual escape kits for the exit party – civilian clothes, false papers, and such things as compressed food, with suitcases or rucksacks to carry it, specifically small enough to be pushed ahead easily by each escaper crawling along the tunnel on the break-out night.

As always, when the time for break-out neared, everybody on the exit list began to say to himself: 'Maybe in a few days time I'll be out, maybe in a month or two I'll be home, maybe...' and that was usually when something went wrong. This was no exception.

At dusk one evening, a ferret discovered the entry to the tunnel forty-eight hours before the diggers were to have broken surface with a periscope, to confirm the estimate of how far south the tunnel reached beyond the wire.

Guards were sent into the hut at once. All the officers living in it were evacuated pell-mell into others. Floor boards were pulled up round the stove to expose the shaft in the ground. The guards brought in chairs and placed them round the opening in the floor on watch for the night. The ferrets and other security officials would take over next morning to make a detailed inspection of the tunnel and then to destroy it.

12

Almost Made It

For everyone who had worked and hoped for the tunnel, there followed the usual thunderclap of disappointment for a sickening half hour. Then our own group of six decided that we did not want to let all that work go for nothing without another try.

If the periscope had shown that the tunnel had stretched under or near the trees on the far side of the wire, the break out would have been made the next night. If all the careful measurements were wrong, it would not be by much, and the further digging necessary would have been quickly done. We decided to take a chance and try to crawl along the ground beneath the hut's floor and into the tunnel that same night, guards or no guards.

A rapid talk with Anderson, he with the Committee, green light given, escape kits authorized and taken with us and, after waiting until nearly midnight in our mess for everyone, and especially the guards, to cool off, we embarked on the next adventure.

We slipped out of our hut around midnight into such darkness that we formed a chain of hands, led by Larry, the best digger. We walked quietly but fast over the soft, sandy playing field, steered by his sense of direction, towards the end of the hut which the goons were occupying.

We could see no light from that end of it nor did we see

anything of the lamp the patrol goon carried on dark nights, wandering with his rifle and, more importantly for us, his huge ferocious dog. But the feeling that this sinister couple might be somewhere near kept us uneasy until we reached the back of the hut and wriggled on our bellies under it.

Our plan had been made in rather a hurry. We had to get under the hut towards where the guards would be sitting around planks torn up above the main shaft of the tunnel. We had to slither into the tunnel under their feet without their seeing or hearing us. And we thought we could do so because we knew, and they did not, about the air shaft to the pumping room.

In fact, the air shaft had been the beginning of the whole enterprise. Its entry was well covered in the ground under the hut some distance from where the guards were now sitting. It had been dug sloping downhill until it reached the level which would become the floor of the main tunnel. From there the shaft was excavated up to the stove and forward to the present tunnel face.

We now aimed to crawl downwards along the air shaft into the pump room. We would then creep through its passage across the bottom of the main shaft, into the tunnel and away.

We could not use the air pump as the guards would hear it. So once in the tunnel we would have to crawl with all speed to its face and then attack the roof vertically to cut an air hole to the surface before our oxygen ran out. That done, we should be able to cut a shaft and climb out before the dawn came.

We hoped the exit would be well beyond the wire and near the trees, into which we would have to move with extreme care, one by one: searchlights swept the perimeter and armed guards walked up and down the fences all night long.

The first man out would trail a string and stop behind the first tree, and the next would follow holding the string, and so on to form a cluster. They were to move fast together if anyone was caught behind them on the way out of the tunnel. We had no time to plan in detail what we would do if we got away. But we had been given the standard escape kits, the railway was not far, and we would sort something out in the morning.

Meanwhile, under the hut, we were in deep darkness through which we had to wriggle in the direction of the guarded area, one behind the other, the second officer touching the heels of the first, and so on back so as not to lose contact.

Larry led us as he knew the lay-out better than anyone and was also to be the first digger to attack the roof. Bert was next because of his massive strength if Larry needed to sit on his shoulders, cutting higher into the sand up to the surface. The third man was Jim Hardie as a relief digger, then Mo, then it was my turn, and Billy the New Yorker was last.

The ground on which we lay stretched under the barrack hut was dry, harsh and irregular and we were constantly pressed against it by the planks above us. Larry, who led us famously, was frequently forced by large lumps of dried earth to make detours and back again to the straight line we needed. He navigated that by touching with one of his arms a central floor beam running north to south under the hut.

Eventually a mild glow was to be seen, still quite far ahead. It came from the electric light above the barrack stove and the opening in the boards around which the guards would be sitting. A few more yards and we heard them talking. Larry lay still a couple of minutes. Then Mo's heels I was touching moved forward again. I followed and kept touching them, and Billy the same to me, slowly, slowly. Then I felt the heels I was touching follow the rest of Mo's body, my fingers and wrists after them, down the sloping, squeezing hole of the airshaft in blinding darkness, wriggling down it into claustrophobia. I felt I was being buried alive, it was so tight.

1. ES kitted out for basic flying training in the open cockpit of a Miles Magister, 1941.

2. Our wedding. Behind, on my left,
our Best Man, Tony Garten,
Grenadier Guards.

3. How Mary sent me Photograph 2.

4. Engaged.

5. Cutting the cake.

6. At 42 Operational Training Unit, Andover, 1941 *(ES seated first left)*.

7. Boston III's about to leave on a Channel sweep *(Photo: IWM CH 6185)*.

8. Wing Commander Reggie Reynolds DSO★ DFC, Commanding Officer 139 (Jamaica) Squadron, with his navigator Flight Lieutenant Sismore, who survived 100 sorties and then became a pilot.

9. De Havilland Mosquito IV *(Photo: IWM E [MOS] 884).*

10. Flight Lieutenant W.C.S. Blessing, a good friend in Jamaica Squadron. Amongst his successful daylight missions was the destruction of the strategically important molybdenum mine at Knaben in Norway *(Photo: Sport & General)*.

11. The marshalling yard at Duisburg after a raid on 13 May, 1943, showing extensive damage *(Photo: IWM C3571)*.

12. Colonel von Lindeiner, *Kommandant* of *Stalag Luft III*. (Photo: IWM HU 21009).

13. *Stalag Luft III*, Sagan, under snow. *(Photo: IWM HU 21026).*

14. One of the two photographs taken of me in 'civilian clothes' for my escape documents by a camp technician, at immense personal risk of torture and death.

15. A Goon tower, drawn by a Kriegie whilst in *Stalag Luft III*. *(Photo: IWM HU 21135).*

16. The set of Noel Coward's *Blythe Spirit* which was performed in the North Compound in 1944 – a remarkable piece of work by our fellow Kriegies, built with very limited resources.

17. Bill Manser (standing) and ES photographed by Willy at the entrance to our hide beneath the old car beside which some SS fought their last battle.

18. Home again – the war over. 1945.

Then it loosened and I could just crawl along and was not buried alive after all, and calmed down again. Then there was a less than darkness visible beyond Mo's feet and bottom, the dim residue of the light above the feet of the guards sitting in the hut, diminishing down the main shaft and all but extinct through to the pumping room.

After a few moments our line halted when Larry reached the end of the shaft and slid onto the floor of the pumping room to proceed into the tunnel as planned. We gave him one minute, then moved up for Bert to do the same, stopped again for a minute for Hardie, advanced for Mo's turn. I edged forward as Mo eased himself out and dimly saw him step across the pumping room and into the passage.

The quicker I did the same, I thought, the less time for one of the guards to think about peering down the main shaft. But I forced myself to wait through the agreed interval and then crept through the pumping room to the shadowy floor of the main shaft. It really was the bottom of a pit, the light scarcely reached us, I heard the guards talking above me, I crouched across the sand, slid into the black mouth before me, and started crawling, Billy followed.

Compared with what we had just squeezed through the main tunnel seemed spacious, an easy crawl. The dim light behind us faded fast. Larry lit the first margarine lamp on its shelf in the wall, and lit another some yards later so that Billy behind me pinched out the first – we had no oxygen to spare – and so on for over two hundred feet where Larry and Bert reached the tunnel face with the rest of us pressed behind them.

Larry, half standing, at once lit the lamp which had been left in the wall near the tunnel face and stripped to the waist. He grabbed a spear-like tool which was leaning there and attacked the roof of the tunnel at an angle just less than vertical, say eighty degrees forward. Bert was on all fours behind him, clawing the falling sand back between his knees like a big black digging dog, the rest of us following suit, Billy and I turning round every few minutes to crawl and smooth the sand back down the tunnel to maintain a thin level.

It has been a noble painting in the gallery in my head every time I have thought of it since then – the daubed and dappled black and orange flickering from the glow of the light on the clean sand close about us, sinewy Larry, Tit fer tat, 'at Larry, in the heroic mould, thrusting and twisting, stabbing up and pulling back at his spear, and stabbing again, gaining height, inch by inch, the fine sand dribbling down to Bert, a burning night scene by Delacroix, fury of muscles, shining sweat, and a tension slowly beginning.

Larry, stabbing and tugging, stabbing and tugging, has climbed to sit on Bert's shoulders, Jim taking Bert's place as digging dog to shove the sand back to the rest of us.

Quarters of hours dragged by. We began to feel tired. Larry dug higher, stood on Bert's shoulders, the sand poured down. We shoved it away from them, someone found a wooden structure for Bert to stand on with Larry standing on his shoulders, the sand endlessly pouring. We felt sure that only three feet above our heads we would get air. And if we got air we would carefully open a hole big enough for a man's shoulders. If we were in the trees, we would slither one by one and away into the forest at once. Another fifteen minutes went by. It was not fatigue but incipient asphyxia which was assailing us.

We were deeper below ground than we had thought. How much higher was the sweet air? If we turned back we could still make it home to the main shaft. Should we play all or nothing? Thus our brief conclave.

Whatever others might have thought about us – and we were far from universally approved – we were disciplined young officers. Our heads began to swim, Bert, still standing upright, colossus to Larry, agreed with the rest of us on all fours, we must pack it in. I saw Larry's shoulders slump in disgust, he climbed down from Bert, we all about-faced, left our escape kits as they lay, crawled, weakening by the minute, back towards the main shaft, Billy first, me second and so on, with poor Larry last.

Three quarters of our way back there had been a fall of sand from the roof since we passed forward which blocked the tunnel two thirds of the way to its ceiling. It was for Billy and me to clear. If this had happened near the face where the air was fouler we would by now have been too weak to do it, and all dead men. But the air was slightly better near the main shaft, we were weak but not too weak, and clawed the blockage with our hands like sleepy cats, enough to complete our crawl to the floor we had to cross.

At the tunnel mouth we again heard the guards talking above us. Billy crept across the pit floor and into the tight passage to the pumping room and the air shaft. I did the same, a few steps behind. But as I squeezed in there was uproar from the main shaft which meant Mo had been caught.

Happy to hear no shooting, I crawled and wriggled after Billy, and eventually up through the trap and out of that unforgettable drainpipe made of unreliable sand. On our bellies again on the lumpy ground under the disobliging planks of the hut we slumped and rested in the pitch black silence for a good half hour.

We could hear no more noise. Later we learned that a guard had seen or heard Mo in the pit and presented his rifle for him to put up his hands and climb up the ladder, and then another guard climbed down it and shone his torch down the tunnel mouth shouting: 'Come back or I fire.' Our four friends were escorted triumphantly to the cooler by the guards who were complimented for having intercepted these would-be escapers climbing into the tunnel.

Billy and I recovered smartly from our trauma when we saw a light along the left side of the space made by the low piles which supported the hut we were under. A light moved up, alongside and past us as we lay scarcely breathing. It was certainly the electric torch of the patrol. I do not know if I really heard, or only projected, the padding of his huge dog's paws, but its presence was much in my mind for a few moments, and then the light went away.

We were under the hut for the long time it took to crawl back the way we had come. At last we wriggled out to the open playing field. The night was lighter. With our ears pricked for the patrol man and his dog, who did not come, we ran north up the field to the main cluster of huts, opened the door of the first we came to, entered into the first mess there, explained ourselves, and rolled up and slept on the floor, disgusted by our failure, tired and unhappy.

Next day, when the ferrets explored the tunnel and right at the end of it found six kits by its face, the inefficiency of the guards was revealed in all its sloppiness and they were punished.

13

Recurrent Thoughts

The next night, dozing on my own straw filled palliasse, I suddenly felt my head was again leading my body on hands and knees behind Billy, back through the dark tunnel, dizzy from lack of oxygen. Time passed. Then again our flashlight showed the massive heap before us. I re-enacted Billy's and my struggle, wriggling over the high sand, our heads swimming, we forced our bodies into the clear air on the other side, we crept like mice up to the tunnel exit, the Germans sitting above and around it.

This was no dream. For years now I have known it was the second of a few such visions in my life, always between wakefulness and sleep, rapidly absorbing the drama, bold in colour and piercing in delineation, each with its particular solemnity, none ever to fade. Within the compass of this book, the first and only other was the vision of Johnnie Skinner which I have described earlier – Johnnie standing upright, profiled in a pool of orange light in my dark bedroom. He was dressed in full flying kit, a Mae West over his brown leather jacket with his seat parachute pushed ahead of him. He was not standing as a wax work, more my friend, still alive before climbing into his Mosquito, standing as to a bugle.

I never again saw that vision, nor ever dreamed of it. Indeed I had no recollection of any dream about the war during or after it, except when I was on the run during my evasion in Holland and Belgium: several times I dreamed I had come home again and was full of joy. On waking what remained for a while were the pictures. What was added was a sorrow.

In the morning I thought I might be getting obsessive about escaping. So I decided to calm down for a bit, at least while our friends were in the cooler, which we had been told would be for three weeks. I spent that day working again on the poem I had begun when I was locked up there myself after my wire job. Since I was released I had had no time to continue it because of my duty to converse hour after hour, day after day, with the ferrets and other Germans who came my way and had to be deflected or at least slowed down from walking where the tunnel was being dug. Now, for the time being, no tunnel was being dug and I welcomed the poem again, in fits and starts, with many intermissions throughout the rest of the war.

A few days later, my first private parcel arrived from home via the Red Cross. The package my wife had put together could not have been better chosen. It contained books on the advice of C.S. Lewis, and she enclosed a copious list he had given her in longhand of further books to be ordered or read from the camp library of the Red Cross. He said that if I read them all, 'the rest would just be top dressing' when I worked for my degree after the war. She had packed in plenty of chocolate bars and tins of Balkan Sobranie cigarettes – I was an ardent smoker at the time and esoteric cigarettes had been one of my adolescent affectations. With them came a silver wick lighter from Aspreys, as the Germans would not allow us to have petrol lighters for fear that they might be helpful in tunnels, and a dressing gown of Air Force blue cashmere which was wonderful as the Silesian winter developed, and stayed with me in all vicissitudes until the final escape.

When my brother and I were children our father had hired an ex-sergeant he had known during their own war in the trenches to give boxing lessons to us. This came in handy a few years later at our private school. There were plenty of bullies there. Although none could box as we did, my little brother was injured and taken to hospital, and it was this that had made my father decide to take us out of the school and send me to Germany for a year. I progressed later, won some bouts in inter-school matches at the City of London and was generally pleased with myself.

Now I started boxing lessons once more, from a man I much admired, Sergeant Northrop, an amateur English middle weight champion with a wonderful attacking technique. His father was a big iron master in North England, and it was rumoured that, before he was shot down, he had flown in a raid on Milan which destroyed one of their overseas factories.

Northrop quickly showed me how lucky I was that I had never blundered into a serious fist fight, took all my style to pieces and, in kind lessons for the rest of the year, taught me much of how it is really done.

Poetry, exercise, and renewed creature comforts made for an agreeable fortnight's holiday from escape-think. Then the relapse came. From a clear blue sky I found myself thinking how lucky it had been that the Germans did not catch me with the others on the night of the tunnel. Coming so soon after the wire job, which had caused quite a stir, I should have certainly been classified as a hard case by the Germans.

My two efforts to date had been targets of opportunity. The Germans had added the wire overhang. And I had the luck to get one place in six for a near breakout which it had taken dozens of other officers months to prepare. But pieces of luck ought not to be waited for. The next effort should be a carefully planned initiative. I was coming to believe that for me it would be home or Colditz, the castle where the most recalcitrant officer escapers were sent, Navy, Army and Air Force, and from which it was extremely difficult to escape.

As always, the choices had to be through the gate, over the wire, through the wire, or well under – by a tunnel. As to the gate, the sentries were meant to examine the papers of every person entering the camp and again on exit, and everything on wheels. But some were not bright and others lazy, and now and then a prisoner got out in a lorry full of uniforms and boots for us from the Red Cross, a bread wagon which had dropped part of its load in one compound and moved to another with a passenger deep among its loaves, under a false bottom in a truck full of filthy garbage. On foot, in British or American camps all over Germany, a hopeful officer would seek to bluff his sentry disguised as another guard, as a visiting German officer, an inspector from the Neutral Protecting Powers or Red Cross Representative, while the real one was at the other end of the camp on his mission. Discovery led to no more than a spell in the cooler. Once or twice there was a home run, and anyhow it was good fun.

Quite a different matter was trying to storm the fences with scaling ladders at night after fusing the arc lights, when the guards would let fly with their machine guns as soon as they thought something was amiss. Some people got away, and some were killed by legitimate fire this way in various camps during the war. Almost as dangerous, was creeping on one's belly across the warning zone at night, over what seemed a terrain with a little cover from a plant or two, or bumps in the ground, or bitter weather, armed with home made wire cutters to cut and slither one's way on the ground through the fence, unnoticed by the goons in their sentry towers and the patrolling guards. Again, some got away, and some got shot. Trying to climb over the wire with a bluff like mine in German uniform was a shade safer: if the sentry got suspicious there was a chance he would call for hands up rather than open fire at once, though I must say this seemed more likely when thinking about it later, back in the camp than half way up the wire.

By far the safest way of getting to the other side of the wire was to crawl under it in a tunnel several feet down. In our camp, as in all the others, the number of tunnels which reached breakout was small indeed and, for each one of those, many were discovered by the goons. If we had tunnel experts so had they, based on knowledge of our techniques acquired over four years which they constantly circulated. Nevertheless, when a tunnel was discovered before breakout, the chances were we would not be in it at the time and could dig another without being personally identified, and another, and another until one night . . . And some tunnels had succeeded in some camps, so why not again?

*   *   *

So far in my story it has gone without saying that escape was the prime purpose of the life into which my friends and I had been parachuted. It never occurred to us to question this. Every time an escape was frustrated, the same emotional pattern recurred: disappointment, despair, amusement and, finally, a determination to make another attempt. We were truly dedicated.

King's Regulations were that a British prisoner was to seek to escape if reasonably possible. But no prisoner was ever ordered to escape, nor reproached for not doing so. The duty was a voluntary choice for individuals, and the spirit of the Regulation was observed in that the needs of escapers were given first priority in Air Force camps, save only for the sick.

Kriegie folklore had it that for every fifty prisoners who attempted to get out of the camp, forty nine were stopped – the tunnel was discovered before breakdown, the disguise seen through at the gate – and one got away. And for every fifty who got away, forty-nine were recaptured usually in a few days and only one got home.*

In the face of such statistics it is not surprising that the great majority of Kriegies accepted that they had fought their good fight and would now settle down to long for the end of the war. This did not mean they washed their hands of escape, they just thought it was almost impossible for them to succeed. Most officers were willing to lend a hand to escapers.

* An official report to the Air Ministry after the war on RAF escapes, by Aidan Crawley, not made public until 1985, states that out of all the British Air Force prisoners who tried to escape from permanent camps in Germany, fewer than thirty ever reached Britain or neutral territory.

14

Digging in the Lavender

By the time my friends were nearly out of the cooler, I had convinced myself that a tunnel of our own would be the best answer, and that we should get on without delay. When the Germans discovered a tunnel before its breakout, it was usually because a ferret was poking about and found the entry trap door. It followed that we would have to find a place which no normally wandering, or specifically poking German could possibly see.

Larry, our cockney digger, Bert, the massive Canadian and the others were let out in the evening. We celebrated as well as we could. The consensus was, let's look around, and before nightfall we had an idea.

Next morning, Larry and I went on a walk round the perimeter. Half way down the west side, going south, we stopped at two rectangular wash-houses about a third the size of our barrack huts, and built of concrete. They backed onto the warning rail which I had crossed to climb the wire a few weeks ago, not far from the same place.

Opposite the southern wash-house, and only twenty feet the other side of the wire, a goon barrack hut stood parallel to it in the faded grass. It was one of a cluster of similar huts at this end of the *Kommandantura*, the large area where our keepers had their offices and lived with no barbed wire enclosing them. It would be ideal if we could run a tunnel from the hut, so near to the wire compared with what we had just gone through.

The wash-houses were gloomy, wet and smelly places of freezing showers, of doing one's own laundry, and nearly public group defecation, reluctantly but indispensably visited every day. But all that be damned, they backed onto the warning zone itself. One couldn't walk nearer to the camp wire without the risk of being shot. A tunnel from either wash-house would not need more than twenty yards to get us under the warning area, and about the same again to be under the double fence and the patrolled path beside it.

The floor of each wash-house was of concrete, dark grey and wet in the room where we used to shower and do our laundry. It was drier in a latrine room which lay behind a creosoted wooden partition. Daylight came thinly through the doorway and some small windows on the camp side. I don't remember if there was any electric light, and if there was the Germans were economical with it. The wall opposite the doorway was windowless, a thick grey slab bearing a maze of grey metal pipes and taps and wash bowls, the warning zone being on the other side of it.

The floor of the latrine room had a long opening down the centre. Over it a row of wooden lavatory seats ran the width of the room, partitioned back and sides, open in front. Beneath them was a large cesspit. It was pumped empty by the Germans from time to time up a thick hose into a tank on a cart which two horses pulled away for fertiliser.

Add another few yards to a tunnel from the south of the two wash houses and we would be under the goon hut. It was only used for their office work and showed no lights in the evening. Moreover it was built on short piles, just like our own barracks. So on break out night we could emerge from the tunnel under the floor of their hut, and slither across to its side away from the camp, invisible to the patrol, unlit by the search lights, and up and away. It was beautiful.

We went straight to the latrine bench. At one end of it two officers were seated in their frontless compartments. At the other end we peered down the holes of the seats with a small flashlight and, as I had hoped, the angle of vision did not allow us to see the sides of the cesspit. If its sides were vulnerable we should have a good starting place for a tunnel. What did Larry think?

He agreed. So we had to find out how the sides of the cesspit were composed. Accordingly, the same evening we sought permission from Anderson to make a reconnaissance inside the tank. Next morning he said the Committee agreed we might do so and report back. They would provide lookouts to saunter in the neighbourhood and send warnings down to us if goons came near.

Armed with that authority, after the morning *Appel* Larry and I carried some shirts in the wooden washtub of our mess into the south wash-house, and Billy followed us. We left our tub in the washroom, and in the latrine room we removed part of a lavatory seat, careful to be able to put it back without trace. I then took off all my clothes. Billy had made me a mask of a handkerchief with tapes sewn to it which I soaked with iodine and tied round my face. He and Larry lowered me by my arms through the lavatory seat.

When they let go, I dropped about a foot, and the urine mixed with shit was above my waist. When he heard what we were going to do the Senior British Officer, Group Captain 'China Bull' MacDonald, lent me a precious, large electric torch. I now switched it on, and found I was in a tank extending below part of the floor of the latrine room which formed its roof. I could just stand upright. The floor under my bare feet was made of concrete and felt slippery. I waded across it into blackness, holding the torch before me, steering a course at right angles from the row of dim-lit, empty lavatory holes overhead.

Turds brushed against my naked thighs and stomach like gentle little fish.

A few steps on, and the side wall of the tank seemed dimly visible, rising above the smooth, black urine on which the torch light danced. But another step, and I saw that at eye level between me and the wall there was an arch parallel to it which supported the roof. It was made of brick and a foot thick. It left only a foot and a half between its underside and the surface of the foul pond I was in. Another two steps and I bent my knees until the sewage came up to my throat, waddled under the archway with the torch pressed against my head to keep it dry, and stood up to see the wall itself a yard away.

It was also made of brick instead of metal or concrete as it might have been. We should be able to hammer through it fairly easily, and the chances were there would be packed sand on the other side under the continuing floor of the washroom.

Apart from the stench the situation seemed perfect. Unless a goon actually did what I was doing – was lowered into the cesspit and went for a disgusting paddle and under the arch – our tunnel mouth could not be discovered. Moreover, we would make a trap to cover the entry hole whenever we were not working, with a face of false bricks reconstituted from the debris.

I shone the torch on myself and shouted 'Larry' through the mask.

'Yeah?'

'Look down the hole. Can you see me?'

'No,' said Larry.

'Can you see this?' I asked, waving the lamp about.

'Just a little,' he said.

The atmosphere made breathing to talk unpleasant. I turned, ducked under the arch and plodded back. When I was under the seat Billy lowered a piece of rope for me. He went through the partition to the washroom and the door to check with the look-outs that all was clear, and I pulled myself up into the washroom.

I must have grown slightly accustomed to the atmosphere below and, stupidly, was embarrassed by the comments of my assistants when I came near them on the way to the shower room.

Billy reassembled the lavatory seat. No one could see it had been tampered with. I scrubbed myself all over in the icy water and we went back to our barrack hut to talk it over with the others.

We agreed we would have to wait for the tank to be emptied: it would be impractical to stand cutting the wall all day with our bodies covered in shit, and our incessant showering would be seen by too many people for security. But the goons were certain to bring in the sewage carts fairly soon to prevent the tank from overflowing. On past form, only an inch or two of liquid would be left in it. And as soon as the goons finished, we would begin.

It was now for us to clear a full working plan with the Committee. We soon settled it among ourselves, and with Anderson. We would be a team of seven, the six of us from the last effort, and Bill Manser whom we had not been able to include on our tunnel job because his mess then was in a different hut from ours and we simply had no time. Our idea was to dig a tunnel from the cesspit to end beneath the goon hut.

We would cut a hole in the brick wall of the cesspit, shoulder high from the floor, and lug sand and earth from behind it until we had a cave one of us could squeeze into. From that bridge-head we would dig a large chamber underneath the wash house so that when we dug the tunnel, we could dump the sand straight into it.

Instead of going deep down and shelving up again, like the tunnel we nearly suffocated in, we proposed to wait until the

115

winter freeze and then run a rapid tunnel, of say four or five feet, below the icy top soil. Even if a goon walked in the zone and over the tunnel line, or if we had a slippage of sand inside the tunnel, there would be no trace on the frozen surface.

We reckoned that from inside the wash-house, the length of the tunnel would be: under the warning area 30 feet, including the chamber; under the double fence 9 feet; width of night patrol path 11 feet; to the edge and well under the goon hut 20 feet; say 70 feet in all.

We could make a small air pump and line if needed. But we could probably do without fresh air in the tunnel as far as the wire, and then bore small air holes up into the thick tangle of wire between the two fences.

Of course digging the chamber would mean tons of sand which we would have to disperse in the camp unbeknown to the Germans. Anderson said he could give us a general all clear provided one of our group reported progress every day or two. The Committee would provide the indispensable screens of look-outs.

He thought we might need help for the dispersal of sand. The Committee could provide that from the dispersal team they had developed from the previous affair. But it would mean letting another two or three officers escape with us in return for all that dreary work.

We were not keen on this as the smaller a breakout party the greater the chances of all getting clear away from the camp before the escape was discovered. Anderson agreed to wait and see how our team managed. But if there was the smallest doubt about our capacity the Committee would insist on their group intervening to help disposal and joining our party. In the event we managed on our own, as will be seen.

While waiting for the tank to be emptied, there was plenty of work for us to do. With a triangular file I had bought from

116

a guard for fifty cigarettes we notched the blade of a table knife to make a keen, small saw.

A pair of pliers which the Committee lent to us had been acquired the same way months before. We also had a hammer stolen from a German carpenter, who would not have reported the loss because of the punishment he would have been given.

With these tools Billy changed a fixed lavatory seat in the middle of the row of holes to a sliding panel operated by pressure from two hands spread apart in a certain squeeze, and so well concealed that we often needed to count our way along to find the right one.

To make a big chisel to cut through the brickwork, we wrenched a metal rod from a disused cooking stove in a store hut. We heated one end of it red hot in the charcoal stove in our hut, hammered a chisel head and tempered it in one of our metal water jugs.

Two thinner bars were also torn from the stove and beaten at one end into hooks and at the other into rings. To these we tied ropes taken from kit bags, and into the ropes we fixed wooden cross pieces to make a rope ladder.

Last, with pieces of wood from the roofing of our hut, a few bed-boards from our beds, and three pairs of hinges from various furniture in the camp we built a folding trestle for the tunnel-diggers to crawl along in the cesspit, safe from the filth on the floor.

None of this was easy. We were in constant danger of confiscation: time after time, a wandering German would appear, the warning from our look-outs would be given and our precious materials would be bundled away.

While so waiting, a tall Australian Flight Lieutenant wearing dark blue battle dress – I barely knew him and have forgotten his name – asked me to come on a walk round the perimeter. It was a golden, sunny afternoon. He said the Germans had ordered fifteen British officers to be transferred from our compound to the North Compound the next day, and he was one. He said a big tunnel was being built there, and an invitation had come for me to join the operation, if I wanted to. My name would be put on the transfer and, with my German, I would have a place among the first to climb out when the time came.

This was a considerable honour, and I wondered if it had come from Squadron Leader Bushell who had discussed the wire job with me. I would have accepted eagerly but for our own project which I thought was even better – a small team, the shortest possible tunnel, and not only an invisible entry trap but an exit invisible to guards and search-lights. I thanked him warmly and said I was already engaged in an authorized project in our own camp. We parted, and he wished me luck, and I the same to him.

A couple of weeks went by. Then one morning while we were being counted in *Appel*, we saw the gate of the camp open, and the horses and two sewage carts dragged in. Two goons barred both wash houses against us and settled down to watch a couple of Russian prisoners pump. We had finished our preparations. By lunch-time the tanks were empty. We would enter ours on the following morning.

That afternoon, the Committee's look-outs placed them-selves in distant, medium, and danger semi-circles converging on the wash house, and practised serial signing for the man on the door to dart in and stop any hammering from being overheard, if goons were wandering our way.

Next morning, after *Appel*, they went to their places and Larry, Billy and I went to the latrine. The tank had been drained to barely ankle depth. Larry and I wore masks soaked in iodine. Billy lowered the rope ladder for me. I climbed down barefoot, my trousers rolled up to my knees. The stink was strong and acrid as before.

Standing in the mess I took the trestle's parts which Billy passed down and put it together. Larry swung down and crawled on it to the brick face of the wall while I paddled alongside carrying a burning margarine lamp.

Because of the moisture to which it had been exposed the cement binding the bricks together was too hard for our home made chisel to cut. So we attacked the brick itself, taking turns every few minutes to hammer on the chisel. The noise was deafening in the confined space of the tank, but no goon came within a hundred yards of the wash house, and with three hours of non-stop hammering we had cut a cone through the bricks sufficient to make a small hole in the far side. The brickwork was eleven inches thick.

Then the doorman yelled his message down, and we had to stand shivering in the smell for half an hour until the 'all-clear' was given.

We resumed our hammering. By the end of the afternoon, we could push an arm through a hole big enough to scoop out some soft sand and rubble from the foundations.

We stopped an hour before the afternoon *Appel* to be sure not to be late for it. Larry crawled back along the trestle and climbed up and out to Billy who stood guard by the lavatory seat. I stowed the rope ladder in the hole we had made, folded the trestle and lugged it on my back, flattened it against the wall, invisible from above, and finished my slippery, bare foot paddling for the day. Billy pulled me up by the arms and I spent plenty of time washing.

We had worked from half past nine to five o'clock and were pretty tired. But we had got off to a good start and felt duly cheerful.

When we got back to our rooms that evening, we had heartening news. Three officers from the neighbouring East Compound had escaped to Sweden after a four-month operation, now justly famous for its boldness and ingenuity.

15

The Trojan Horse Escape

In the left-hand corner of the East Compound there was an extension, the Canteen, some forty feet before the trip-wire. It was constructed on a red brick foundation with a flight of wide wooden steps and contained the cookhouse, a theatre, a barber's shop, a music practice room and, in the evening, a storage room for the Kriegies' sports material for football, basketball, golf, boxing, and so on.

One day in June, 1943, the footballers and a few goons were surprised to see four strong Kriegies carrying from the Canteen a big rectangular object on wooden poles and set it down on the open sand, about as near to the trip-wire as the start of our own tunnel. The poles were withdrawn and the men began vaulting over the box. Soon others joined in. One man was particularly clumsy. The guards laughed whenever he fell over having made his try – especially the one in the near-by watch tower who had nothing else to do. The more they laughed the more he tried. At last he fell over and the vaulting horse lurched on its side and the guards could see its empty inside.

Two hours later the vaulters stopped and carried the horse back into the Canteen. Next day the same performance was carried on using exactly the same spot. As days went by, the goons got used to it, and more Kriegies joined the vaulters.

Two weeks earlier, Flight Lieutenant Eric E. Williams had

obtained the agreement of the Escape Committee to build what was to be known as the Trojan Horse and use its real activity to dig from under it a tunnel going under the barbed wire, into the forest, and away. Reversing the strategy that took Troy, the scheme solved two of the problems that bedevilled Kriegie tunnellers – the concealment of the entrance and the distance to the wire. Lieutenant R.M.C. Codner of the Royal Artillery and, later, Flight Lieutenant Oliver S.C. Philpot were with him. The operation began on 8 July, 1943.

Inside the horse's frame there were cross-pieces to allow two men to squeeze themselves in and a row of bags sewn from trouser-legs were hooked to one of the bars to carry out the sand.

Some officers had formed an orchestra, rehearsing music by the window of the Canteen, and whenever a ferret approached the violin stopped playing. The danger gone, it resumed its mournful tune.

After the first week of vaulting, Codner was one of the four men holding one end of the pole as they carried out the horse. Williams crouched inside it, his feet on the framework. His arms held the equipment: a cardboard Red Cross Box for the surface sand, the bags and hooks, one side of the vertical shoring for the tunnel and a trowel. The Kriegies put the horse into its usual position. Underneath the horse, Williams scraped up the dark surface sand, put it into the cardboard box and started to dig a vertical trench for one side of the shoring, emptying the sand into the trouser-leg bags. A few days more and he started digging horizontally. The hole was covered by a trap of bed planks and the surface sand carefully replaced at the end of each shift, so that it was not distinguishable from the much trodden sand of the vaulting surface.

The tunnel was three-quarters dug when, without warning, ferrets, escorted by armed SS on motor cycles, launched a dawn raid. They went right up to the canteen where the ferrets immediately began a ferocious search. The sports gear stood on one side. The ferrets found inordinate amounts of yellow sand in the roof and under the bricks. It had to be taken away in trucks. Colonel von Lindeiner himself turned up. Then they found a tunnel in another part of the camp, and the panic was over.

Two or three days after this alert, the digging was resumed.

To be cooler, and to avoid extra sand being scraped from the sides of the tunnel as he moved backward with the sand, Williams worked naked. Sand was in his eyes, ears, nose and under his foreskin. After 40 feet of digging and two months in the dark tunnel, with insufficient air, insufficient food, and overworked, he was taken into hospital suffering from exhaustion. He stayed there a week during which no digging was done, although the jumping continued.

Williams found a guard they called 'Dopey' who was chatty and not so bright. Dopey accepted bribes of chocolate and cigarettes in exchange for all sorts of information, and liked to moan about his wounds from the Russian front, where he had been left lying in the snow; his greatest fear was to be sent back again. Williams wanted to hear how often the trains passed through Sagan, about the *Arbeitskarte** that foreign workers carried and the special police passes they needed to travel by train. Lots more chocolate and cigarettes changed hands and Dopey showed the papers. When asked to leave them in the camp for a day the guard was terrified and refused. But Williams said the bribery would be revealed: 'We will be sent to the cooler for two weeks, but you – you will be shot'. Dopey gave in. Careful copies of the passes were made before returning them, and the train schedule studied to make sure it would not change before the end of October.

* Work permits.

After Williams returned from hospital, Codner also went into the hole, a metal basin like a toboggan with a rope at each end tied to his ankle to more effectively drag out the excavated sand toward Philpot, who also took care of spreading it among the other prisoners so that a human chain was disposing the bright sand brought out from the tunnel.

Near the end of October, the tunnel was ready, almost one hundred feet in length. The exit was to be just beyond the patch patrolled by the sentries outside the wire, but it would be necessary for the escapers to cross the road and reach the wood on the other side unobserved. There was a short period, after dusk, before the night sentries came on to walk all night long around the outside edge of the camp.

Williams and Codner broke surface when it was nearly dark, followed by Philpot. Their black camouflage was discarded under the trees, revealing 'civilian' clothes that had been re-cut, re-dyed and re-fashioned out of uniforms and blankets. Of course they carried some escape food and forged identity papers.

Williams and Codner were going to Stettin together, Philpot, alone. They split up and walked to Sagan railway station. As they approached their train, they ran into the camp doctor. Fortunately Williams had shaved his heavy moustache and the doctor passed without recognizing him. After Frankfurt (on the Oder) and Küstrin they arrived in Stettin, where they slept the first night in the air-raid shelter of a suburb garden. They spent a few days looking around the docks, changing hotels, passing their afternoons in a cinema – a safe place – before starting their nightly rounds of pubs where they hoped to meet Frenchmen to help them get out of Germany. One Frenchman, who was escaping himself, helped them get on board a Danish ship which made route to Copenhagen and Oslo. The fugitives remained below deck for four days. Finally, one of the ship's officers put them on a Swedish pilot ship to Gothenburg from where they reached the British Legation in Stockholm. A few weeks later they were home.

Philpot, who was posing as a Norwegian, had decided to travel to Danzig alone. He also joined the train to Küstrin from where he took the express northeast to Danzig. He had excellent papers, good clothes and took great care with his appearance, wearing a Hitler moustache, smoking a pipe to account for his mumbling (he knew no Norwegian), and pretending to be on a tour of branch factories for his company. He reached Danzig safely after only one identity check on the train and survived a night sharing a hotel bedroom with a German. The next day he joined a happy crowd of carefree tourists on a boat trip around the harbour, ignoring a sign that said: 'Passengers who take these boats are warned that they must have special permits to enter the dock area.' There was no check. His eyes searched the big and small boats. Docked there he discovered some skilfully hidden German submarines and then, suddenly, the large yellow and blue colours of the Swedish flag floating in the wind, and round the stern the letters: *Aralizz*, Stockholm. She was loading coal. In his book, *Stolen Journey*, Philpot recalls his excitement. Here was his ship, and his plan was simple: sneak into the forbidden area and climb along the mooring ropes through the stern hole into the ship.

That evening he made his way by tram and foot to the distant Swedish dock near the mouth of the Vistula where the *Aralizz* was moored.

As darkness fell, he got to the quay-front from an adjoining landing stage. He crept along the fenders of the sea wall, out of sight of the sentries on the well-lit dock above, and towards the two stern cables looping up to the hull of the ship.

Then a boat approached, forcing him up onto the exposed dock. Desperately seeking cover, he dodged the sentries before making a dash back for the bollard holding the two cables. It was only when he had climbed painfully to the top that he saw the cable disappear around the far side of the ship, pressed tight against the hull. He had no choice but to go back down, and up the other cable.

Philpot at last hauled himself exhausted onto the poop of the *Aralizz* to meet a terrified absence of welcome from the crew. The captain ordered him off the ship. Philpot pleaded, offered money, then refused to move.

The captain walked away. The *Aralizz* was due to sail in two days. She would certainly be searched before leaving and most likely intercepted before she was out of territorial waters. Philpot tried burrowing in the coal, but realized that would not hide him from guards with dogs. Then a Swedish sailor bolted him in a tank at the bottom of the ship. Supplied with food and water, he spent twenty-eight hours in the tank before the iron trap-door was unbolted and he could emerge to freedom. The captain was the first to shake his hand. For the rest of the journey, to Södertälje, just south of Stockholm, he was treated like an honoured guest.

By the time we broke through the cesspit wall, Philpot was enjoying the hospitality of the British Legation, Stockholm.

16

More Lavender

We talked about this marvellous news till late in the night, feeling happy and elated. Next morning we went back to our digging.

We had scattered the brick and concrete chips and our handfuls of sand onto a wooden box on the floor, inverted to provide a mounting block. But from now on we would have to dispose of the sand. And the cesspit floor could not be used for that. The ferrets sounded the levels of the liquid periodically. They were well aware of the average rate of rise from what may be styled normal sources, and well knew that any sharper rise called for nets on poles to see if sand was being poured in, a sure sign that a tunnel was being dug somewhere in camp.

As already described, a metre below our compound there was grey top soil and dirty sand, then close packed virgin sand descending deeply. When dug loose, it increased considerably in volume and was a highly visible yellow. Whenever a ferret caught a glimpse of fresh sand where it shouldn't be, he reported it and German security would at once be tightened all over the camp. And if they grew more jumpy, sporadic searches of our huts would be made, thorough ones with armed troops moving in along with the ferrets. We were barred from entering our huts while they spent hours lifting planks, taking furniture to pieces, going

126

through our meagre goods, our poor burglars, their eyes popping for anything which years of thwarting us had taught them to look for.

But, most happily for us, the very day we started hammering the Germans gave permission for some officers to dig a refuse pit by their barracks not far from the wash house. And our fresh sand would be easy to mix with their legitimate pile as it grew.

Over the next two days we hammered a hole through the bricks behind the arch in the cesspit, enough to admit the head and shoulders of one of us. As Larry and I stood by the wall of the tank, widening and deepening the cavity on the other side of the bricks, the one who was not digging ferried water jugs, two by two and packed with sand, below the entry seat. A hook would be lowered by Billy and up went the full jugs and down came the empties. Others of our team would take the jugs from Billy, and if the area was 'all clear', would leave the wash room and empty them in the pit.

Soon the cavity behind the bricks was big enough for one of us to start digging inside it and pass the full jugs out. Soon again, the two of us were jammed inside and a third took over the ferrying. He had much improved the standard of living by scrounging a pair of black gumboots from a guard for a large parcel of cigarettes, and so ended the barefoot paddling which had been my bane.

As the chamber grew larger we needed three in the cave, one cutting the hard fresh sand with a spade, another crouched behind him pushing the loose sand back, our dog digging trick, the third filling jugs and passing them through the trap.

By the time the work on the refuse pit was finished, we had enlarged our chamber to some five feet high, six across and six along the line which the tunnel would take. The concrete ceiling above us and the brick wall in which we had cut our way were bare of sand. And all of it been safely swallowed

with the sand from the refuse pit, which from time to time
we were glad to see was carted away under German supervi-
sion to be dumped somewhere in the Vorlager.

Moreover, just as the work on the pit was finishing a
happy quirk of architecture allowed us to stow sand inside
the tank itself. The roof arch and our wall were so close
together that we were able to make a scaffolding with a
plank propped high up against the arch and its other end
resting on the top of a bed board standing up against the
wall (see Figs 2 and 3)

As soon as we could push in no more sand another plank
and its board were placed alongside to be filled with sand
parcels. And so we proceeded, elongating the chamber in the
direction the tunnel would take, and all the time stacking the
cartons of sand in the ceiling.

We finished the chamber near the end of November. It

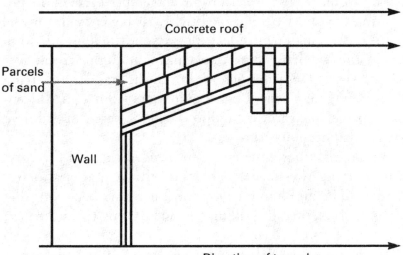

Fig. 2: And across the plank we piled oblong Red Cross
food cartons stuffed with sand, tight and snug against the
ceiling:

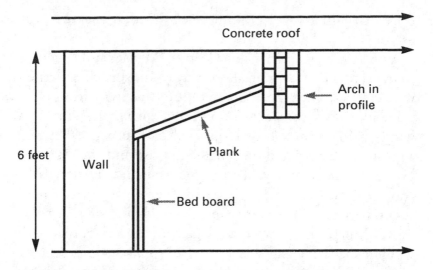

Fig. 3: As soon as we could push in no more sand, another plank and its board were placed alongside to be filled with sand parcels. And so we proceeded, elongating the chamber.

stretched twelve feet long to almost underneath the foundation of the wash house wall on the warning zone. There we sank a four foot shaft in which we would stand to begin our tunnel when the freeze came.

By now the Committee was pleased with our progress and had put in hand the preparation for the journeys we hoped to make after break-out. Their specialists, both NCOs and officers, created our escape equipment. Maps were painstakingly transferred from the few originals which had been obtained from corrupted guards and were in the safekeeping of the Senior British Officer. Travel papers were forged by the experts which each camp seemed to possess: men who had been advertising artists, or draughtsmen in civil life. Compasses were made from magnetized razor blades, the compass cases of a melted gramophone record, the glass cut from the fragments of a deliberately broken window. Money from guards was brought in for cigarettes in steady, small doses to add to the Committee's hidden treasure for escapers.

With our fluent French and German, Manser and I decided to travel together by train all the way to Innsbruck as French workers and, from there, go by foot to the Swiss frontier.

If anyone will consider the minimum he has with him on a winter train journey abroad – his civilian trousers, jacket, and a greatcoat, a shirt and tie, gloves; suitcase and wallet; in his pockets, letters, money, identity papers: in his suitcase, bathroom items, pyjamas, at least another shirt, spare linen, socks, perhaps a few more papers – and will remember that most of these things were not legally in the camp, that each was indispensable and that many had to be made in secrecy – then it will be understood how difficult it was to prepare for our train journey.

As always the Committee performed miracles. Pale fawn cloth caps were made from American officers' trousers, cut to pattern and sewn, with cardboard-lined peaks. Our suits were of American GI tunics and trousers, dyed, as usual, from the leads of indelible pencils which had been crushed and mixed with salt. Their military buttons, flaps and pockets were removed, the lapels altered, the front of the coat cut like a single-breasted jacket, a breast-pocket sewn in and trouser ends turned up. We then had two tidily single-breasted suits, coloured chocolate brown with a purplish tint.

RAF airmen's greatcoats looked like civilian coats when their brass buttons were removed and black ones sewn on. For this purpose, the owners of twelve coats gave up the black inside button of each, a valuable sacrifice in our tight conditions. Even letters in French were written for us to carry in our pockets as additional proof of our identity.

All the above was for Manser and myself. Further outfits for the other five, depending on how they intended to travel, were created, day by arduous day, by the clothing team and stored by the Committee, none of us knew where. It was to

130

be issued on the big night and worn, or packed with our escape food into suitcases or rucksacks which, like those made for our last attempt, had strictly to be small enough to be pushed easily ahead in the tunnel. On top we would be wearing thin black dungarees to protect our escape clothes from the sand and general mess of crawling along the tunnel and wriggling our way under the goon hut. And we would have thin black gangster hoods in our pockets to put over our heads as we left the tunnel.

Once the chamber and shaft were finished, everyone in our team was relieved. Most of the drudgery was behind us. The huge volume of sand was snugly stacked against the roof of the tank, all our other preparations were made, there was nothing to be seen by the outside world but our comings and goings at the wash house among many other officers for the obvious reasons, morning and afternoon. Now all we had to do was to wait on the weather and then get digging a tunnel of only twenty five yards to a point under the goons, and go.

Being able to dump all the sand from the tunnel into the chamber without any of the usual arduous and, above all, risky dispersal, was going to save an enormous amount of time and effort. We would have one digger per shift at the tunnel face, lit by one only of our flickering margarine lamps. A loader behind him would shovel the sand into shallow metal wash bowls. These were fixed by a rope round a stake so that as a third man at the mouth of the tunnel hauled them back to tip their loads empty bowls came along for more.

We thought seven days should do it easily, once the freeze came. The job included cutting a shaft at the end of the tunnel, which would be wide enough for two men to stand in. There would be a ladder up one side and, as soon as we broke earth at the top of the shaft, two of us would wriggle into the gap between the earth and the floor of the goon hut on its piles. After a pause for breath, and to listen, we would worm our way to the side of the hut away from the camp, pushing aside any lumps of earth or bits of wood and stones so that on break-out night we could slide easily across and away.

While this gardening was going on, half a dozen short planks, which we had already measured and cut in the chamber, would be pulled along the tunnel to the shaft and screwed together, there to make a roof for it. Its upper side was camouflaged with rough earth and pieces of wood glued on, and there was a trap door in it which opened downwards.

On the night, one man at a time would creep up the ladder through the trap and, on his belly, under the hut. A member of the Committee, who would not be escaping, would stand in the side of the shaft and despatch them under his orders.

It was agreed, according to common practice, that as the initiator I should go out first with my partner, Manser, close behind me. Where I slithered from under the hut I would spread a ground sheet. Each of the others who followed must go over it on hands and knees to avoid leaving traces on the ground by the edge of the hut. The last man must take it with him and push it back a yard with his feet.

We would fix one end of a strong cord to the top of the ladder. I would carry it trailing so that by running it through their fingers the others could join Manser and me at our rendezvous in the dark. This would be a path between two other huts some forty yards further from the camp.

I would carry a second cord fixed to the ladder, which the Committee man would hold and, according to jerks from me, despatch our followers. When we were all together by the two huts, I would give the farewell jerk and the Committee man would pull in the cords. We would then move in single file across a stretch of shrubbed garden to a few trees which we could glimpse from the camp, cover another hundred yards or so and step over a low garden rail of the *Kommandantura* with a public road on the other side. Then into the forest and separate according to our different partnerships and to the routes vetted by the Committee.

For his part, the Committee man would close the trap from below and return to the tank. There he would wait until morning to emerge with other prisoners using the latrine and wash rooms.

We also had an alarm signal in case any night guard caught a man before the rendezvous. In that case, the Committee man would close the trap and retreat to the cesspit with any others still under the hut or in the tunnel.

Unless any of us were caught during the escape the Committee would try to cover our absence in the *Appels*. Experience showed this might work for a day or two but not for long. But even when the Germans discovered we were missing, they would not necessarily find the tunnel at once. In that case, the Committee would hope to use it for a few further escapes, including the Committee man who had ushered us out.

We drifted into December. A little snow fell. The sun shone next day and melted it. Then came frowning weather; in four days the top ground was frozen. Within ten days or less, we would make our 'Blitz tunnel' and be on our way, we thought, happily testing the ground with our heels.

17

Belaria

The following morning, climbing out of our bunks we were told to pack up. The Germans were moving all British officers out of the Centre Compound into another camp, Belaria.

I fought back the shock and empty feeling in the pit of my stomach. The others swore or mourned. However, we were all hardened to such blows.

Only Billy, our American digger, was left alone for our marvellous project. As a member of the US Air Force he had to remain in the Centre Compound, and make an immediate report to the commanding American officer.

Belaria was newly built, six kilometres from the rest of *Stalag Luft III*. It was under the same *Kommandant* with a sub-commander living nearby. We were the first Kriegies to inhabit it. The bustle of settling into a new camp helped us past the residual gloom of having lost our Blitz tunnel.

Billy was, of course, in the Centre Compound and Larry, RAF, our cockney digger, preferred to mess with some other friends. That left five of our dedicated clique together: Jim Hardie, RAF, a Texan volunteer; Bert Larson, Royal Canadian Air Force (RCAF); 'Mo' Macay, RCAF; Bill Manser, RAF; and myself.

To make up our number, as ordained by the Germans, our hut commander posted to us three officers whom we had not previously known, red-haired Terry Field, RCAF, another now forgotten, and yet another Canadian named Art, tall, well built, a film star kind of face, golden hair, pale blue eyes and lacking all charm.

We were now generally more comfortable than we had been in the Centre Compound. Our new home was a small rectangular room, the sides and roof of new pine wood, still pale yellow, and designed for eight. After darkness fell it was lit by bare electric bulbs, hanging from the ceiling. The entry was through a door from the corridor which ran down the middle of our barrack hut. On the opposite side of the room was the outside wall of thicker wood, with two glass windows and wooden shutters. These had to be closed at specified night times under German orders.

We had a bare wooden table and two benches, so that eight could sit at the same time, and low cupboards under the windows. Four double bunk beds along the three walls other than the window wall completed our furniture. On them, another blessing, instead of the straw palliasses of our last billet, each had a passable stuffed mattress.

Belaria stood in much more open surroundings than our Centre Compound. There our vistas had been limited by dense forest and dreary administration huts, while our view east gave onto the East Compound, almost a mirror image of our own. Here in Belaria our perimeter promenade enclosed most of the goon buildings on one side and, on the other, we could gaze across the wire on open farmland stretching away in low billowy hills with well spaced clumps of deciduous trees in the snow, black traceries on white.

It was a pleasant change to see wide distance again. However, there was a price to be paid, for our new perimeter had no trees or even bushes near it to help hide exit traps from night patrols. Furthermore, we would have to tunnel through heavy clay with a water table only five feet down, instead of the close-packed sand, dry to a depth of eighty feet, to which we had grown accustomed. Nevertheless, even before Bill Manser and I could think up a plan, a like-minded officer, 'Paddy' Paddock, found a reasonably promising corner under a hut from which to start digging. I remember going half underground into what would be the shaft, and his offer to Bill and I to join his team. We accepted at once.

Meanwhile, word came down from Centre Compound that Billy and a crew had been ready to go out through our lavender tunnel for two nights, but were delayed. A passing ferret had got suspicious and looked under the latrines, lifted a couple of loose blocks and stepped down into three feet of shit. Waving his hand torch, he discovered the sand above his head. Rather bewildered, he pushed upwards and half a ton of sand poured down, nearly squashing him to the ground. The goons were suitably annoyed and we in Belaria had a good laugh.

But soon after, all games stopped. It was the end of March when the Senior British Officer, Group Captain 'China Bull' MacDonald, called us on parade. He stood and faced us all under the appalling sky, and said that fifty officers who had escaped from a tunnel in the North Compound had been murdered by the Gestapo after recapture.

To us it was vast thunder, unimagined, huge as the Day of Wrath.

At the end of his announcement, MacDonald added curtly that all theatricals and sports would cease until a requiem parade and service had been held, and that all escape attempts were forbidden until further notice.

*   *   *

Bit by bit the awful story was revealed. The murdered officers were among several hundred who had been moved to the North Compound when it opened on 1 April, 1943. By 11 April work had begun under 'Wings' Day and Bushell's leadership on tunnel entry traps in three different huts. The three tunnels 'Tom,' 'Dick' and 'Harry' were to be dug simultaneously – Tom and Dick to the west on diverging courses, and Harry to the north, so that if one were discovered the Germans might be lulled into relaxation.

Tom was indeed discovered in the early autumn by a ferret tapping the trap with a pick. By then the tunnel was two hundred and eighty five feet long, its front deep below the

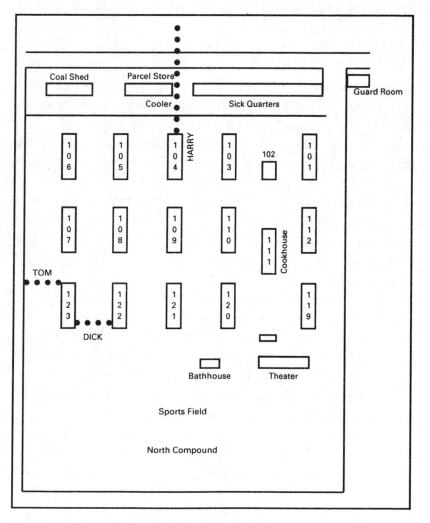

Fig. 4: Tunnels Tom, Dick, and Harry

wire and the first trees. Only a day or two more of digging was needed, before the shaft, the ladder, and the escape trap were constructed and the date of the breakout selected.

At this point Wings and Bushell ordered Dick and Harry closed until the New Year of 1944, some three months away. This was principally to allow the Germans to drop their guard, and also because the Silesian winter was so severe that it was only suitable for escapers with enough German to travel by train.

On 15 January, work resumed on Harry and continued vigorously, and later almost frantically to meet a self-imposed deadline of 24 March – the most propitious night for a breakout because it was moonless. At 9.45 that evening, Harry's escape trap was firmly in place and shut tight.

Aidan Crawley reports that Harry was three hundred and thirty-six feet long, the entrance shaft twenty-eight feet deep, and the exit shaft twenty feet high. On the same epic scale, it was estimated that some six hundred Kriegies had worked on the project, one way or another, and deserved a chance to escape, as did some others whose active camp duties had prevented them from joining Harry as they would otherwise have wished to do.

So, after giving priority to those who had organized and specifically worked the tunnel, a lottery was drawn to bring the total up to two hundred for possible escape, allotted in priority blocks. This was considerably more than a single night's tunnel breakout, one by one, could probably achieve, but was justified by hope of good luck. Full equipment, clothes, maps, and so on were created for them all.

On the 24th, after evening *Appel*, the tunnel was prepared for the great departure. Electric lights were fixed and switched on, the air pumpers got into their kennels, the trolleys on which each escaper would be towed to the exit were equipped with new ropes.

At 10.30 the first man was tugged along. Throughout the

night, the guards in the sentry towers were changed every two hours. The searchlights swept continually by, a patrol guard walked his long beat up and down the path by the wire. In the dark intervals, at the top of the tunnel shaft, an officer controller guided escapers out. They walked with a fixed cord running through their fingers into the trees to join a group of ten who then dispersed in planned directions. Every twenty despatches the controller took his turn and was replaced by another. For hours the passage continued, overcoming inevitable problems: men or luggage got stuck and time was wasted in the struggles, a Royal Air Force raid on Berlin caused the electric lights to go out, and margarine lamps had to be passed up and lit; and so on.

At 4.45 am, 25 March, a guard on patrol did not use the path by the wire as the others had, but walked along the edge of the trees on a line a few inches from the tunnel mouth. It was still quite dark, the escapee who had just left the mouth lay motionless on the ground, the guard made him out, fired a shot in the air, pulled out his torch, and ordered the next man, still standing in the escape shaft, to climb out at rifle point.

This was the very moment when the last escaper to enter the tunnel mouth had been told that no more would be following as dawn was approaching. He stepped back and the few still in the tunnel reversed and crawled back into camp.

Seventy six had escaped.

The disappointed officers filled Harry's barrack hut with smoke from burning their maps and forged papers and hid as much travel equipment as they could before 6.30 a.m., when Colonel von Lindeiner and a riot squad with steel helmets marched up and surrounded the hut with heavy machine guns. The doors were opened, the prisoners were called out, made to strip naked in the falling snow, searched and sent to other huts, and Harry's was closed down for several days. Oddly enough, the Germans failed to search the hut and when the officers were allowed back, they still had a treasury of hidden escape material.

The day after, 26 March, Colonel von Lindeiner was relieved of his duty as camp *Kommandant*.

As usual, hundreds of thousands of Germans were sent out all over the country to recapture our officers. Most of those who had decided to rough it out in the nearby countryside were recaptured in the first day or two; others in railway trains when police suspected an accent; some right on the Swiss border which they did not approach carefully enough, or in a Baltic port when they failed to find a helpful Swedish sailor. All fell to the erosive ill luck which was the dark side of our Great Game.

Until now, the *Luftwaffe* had always received recaptured RAF escapers from the police to face nominal internment in the camp cooler. And then it was on with the game for a hoped-for home run.

Now there was a sea change. This time the news reached Hitler, Himmler, Keitel, Goering & Co. who, as the spiteful gutter boys they always were against the helpless, overruled the brave disapproval of at least one German general with them at the time, and ordered that all of the recaptured escapers were to be handed over to the Gestapo. Fifty would be shot.

And so the terrible harvest proceeded. Hitler's fifty were murdered by the Gestapo after some miserable days of idiotic questioning and virtual starvation in icy, filthy cells, taken in couples at night in cars, driven to the edges of dark woods, told to get out for a pee in the grass and shot in the back.

Of the seventy-six who escaped from the tunnel, two Norwegians in the RAF reached England via Sweden, and one Dutch officer via Breslau, Holland, and Spain. Five, including 'Wings' Day, were sent to Sachsenhausen concentration camp, which they survived, indeed escaped from and were recaptured. The small remainder were later returned to our own camp and to another Air Force camp at Barth.

Later in the day of Group Captain MacDonald's awful announcement, we were ordered to send the Germans to Coventry, this being the strongest condemnation with which we could prudently assail them: we would not return their salute, we would not talk to them and would not reply to them except under threats. It was astonishing how much they took this to heart. We seemed to be lumping them with the murders while in fact – as we well knew – from the *Kommandant* to nearly all but a few strongly Nazi personnel, they were horrified and ashamed of the crime.

In the few days before the memorial service the whole camp was subdued as if waiting for a heavy storm. It burst in the tremendous statement of the requiem sung by our camp choir and the heartbreaking solemnity of the Last Post on 13 April. The parade was dismissed and we began to face the future.

Not so the Germans, who did not grow angry or threaten us for our behaviour. They were truly hurt, said we were unjust, and complained to such a degree that after a week, and some talk between their and our senior officers, we stopped the boycott.

Nor was this mere show play on their parts, as proved by Captain Wolf, the German officer whose duty it was to count us on *Appel* twice each day. We heard that on the day the murder was reported to the camp, he went to the *Kommandant*, said that he could not face the British officers with this disgrace, and asked for a transfer to the Eastern Front. He received it and was killed in action.

* * *

Before the end of April, the German police brought fifty cremation urns to the camp authorities who brought them to the North Compound. Each bore the name of a murdered officer. The murdered prisoners were:

Flight Lieutenant H. Birkland	Royal Canadian Air Force
Flight Lieutenant E.G. Brettell, DFC	Royal Air Force
Flight Lieutenant L.C. Bull, DFC	Royal Air Force
Squadron Leader R.J. Bushell	Royal Air Force
Flight Lieutenant M.J. Casey	Royal Air Force
Squadron Leader J. Catanach, DFC	Royal Australian Air Force
Flight Lieutenant A.G. Christiansen	Royal New Zealand Air Force
Flying Officer D.H. Cochran	Royal Air Force
Squadron Leader I.K.P. Cross, DFC	Royal Air Force
Lieutenant H. Espelid	Royal Norwegian Air Force
Flight Lieutenant B. Evans	Royal Air Force
Lieutenant N. Fuglesang	Royal Norwegian Air Force
Lieutenant J.S. Gouws	South African Air Force
Flight Lieutenant W.J. Grisman	Royal Air Force
Flight Lieutenant A.D.M. Gunn	Royal Air Force
Flight Lieutenant A.H. Hake	Royal Australian Air Force
Flight Lieutenant C.P. Hall	Royal Air Force
Flight Lieutenant A.R. Hayter	Royal Air Force
Flight Lieutenant E.S. Humphreys	Royal Air Force
Flight Lieutenant G.A. Kidder	Royal Canadian Air Force
Flight Lieutenant R.V. Kierath	Royal Australian Air Force
Major A. Kiewnarski (Polish)	Royal Air Force
Squadron Leader T.G. Kirby-Green	Royal Air Force
Flying Officer W. Kolanowski (Polish)	Royal Air Force
Flying Officer S.Z. Krol (Polish)	Royal Air Force
Flight Lieutenant P.W. Langford	Royal Canadian Air Force
Flight Lieutenant T.B. Leigh	Royal Air Force
Flight Lieutenant J.L.R. Long	Royal Air Force
Flight Lieutenant R. Marcinkus (Lithuanian)	Royal Air Force
2nd Lieutenant C.A.N. McGarr	South African Air Force
Flight Lieutenant G.E. McGill	Royal Canadian Air Force
Flight Lieutenant H.J. Milford	Royal Air Force
Flying Officer J.T. Mondschein (Polish)	Royal Air Force

Flying Officer K. Pawluk (Polish)	Royal Air Force
Flying Officer A. Picard (Belgian)	Royal Air Force
Flying Officer J.P.P. Pohe	Royal New Zealand Air Force
Lieutenant B.W.M. Scheidhauer	Free French Air Force
Pilot Officer S. Skanzikas	Royal Hellenic Air Force
Lieutenant R.J. Stevens	South African Air Force
Flying Officer R.C. Stewart	Royal Air Force
Flight Lieutenant J.G. Stower	Royal Air Force
Flight Lieutenant D.O. Street	Royal Air Force
Flight Lieutenant C.D. Swain	Royal Air Force
Flying Officer P. Tobolski (Polish)	Royal Air Force
Flight Lieutenant E. Valenta (Czech)	Royal Air Force
Flight Lieutenant G.W. Walenn	Royal Air Force
Flight Lieutenant J.C. Wernham	Royal Canadian Air Force
Flight Lieutenant G.W. Wiley	Royal Canadian Air Force
Squadron Leader J.E.A. Williams DFC	Royal Air Force
Flight Lieutenant J.F. Williams	Royal Air Force

Less than two months after we buried our murdered comrades, the Allies landed on D-Day, between Cherbourg and Le Havre.

18

Belaria – Rolling On

Given the interdiction on escaping, we had to seek other diversions for the months ahead.

For my part, I sank into much of the English literature received from home on the advice of my tutor, C.S. Lewis, enhanced by the considerable camp library, thanks to the Red Cross. Manser's main consolations too were profound reading and writing. As to our companions, Bert developed gymnastics in a major way, the others made plentiful use of the playing fields, and in the long confinement in our huts after nightfall we and other officers played poker and bridge galore.

We heard of British Army camps where officer prisoners fell into serious debt. Our Senior Officer ordained that at the end of each month all winnings were to be scaled down to a maximum of five pounds for which the losers were to sign chits for settlement after the war, and this sensible order was generally observed.

Bill Manser and I also carried on with boxing, much regretting the loss of Sergeant Northrop, who had been moved from Belaria along with the other British NCOs we had known in the Centre Compound. There was not much enthusiasm in the camp for the sport, and I have only a general memory of agreeable sparring bouts with our few fans, and two less so.

A heavy Frenchman, Jean Regis, a *'nom de guerre'*, managed to land a punch on my left ear so heavy with his big soft glove that I felt the drum pop. To test it I lit a cigarette, pinched my nostrils and blew hard. Sure enough, out came the smoke from my ear. I was just lucky that it mended soon.

The other incident was a reverse. A young American serving in the RAF and I decided to have a bout with the reservation that we would not hit hard. After sparring around for a while, I thought I would try a trick which Northrop had shown me. It was to swing one's upper body imperceptibly to the right while exposing one's chin and, as the partner punched at it, to swing back so he missed and one hit his chin with a straight right.

All went beautifully to plan, but he had swung so hard on to me that my punch knocked him over and for a few seconds out on the ground. I was much embarrassed and he was much annoyed.

Manser and I also started to learn fencing with equipment from the Red Cross, the weapons being under parole. As our fencing master, we had an expert French officer who had got to England at the time of General de Gaulle's declaration in 1940, and had had a distinguished career in one of the Free French squadrons.

There were not many French prisoners with us. By definition they would mostly have been in the Free French Air Force which was a relatively small force of the highest quality. I can only recall Jean Regis, my sparring partner, and Philippe de Citiveaux, then of major's rank, who was shot down in the *Aero-Navale*, and after the war rose to Admiral of the Fleet of France. I remember us doing our laundry side by side in the cold water.

With so much time spent in such close quarters as ours, good manners were vital and vigorously maintained. For instance, talking about the wonderful dishes we would have when the war was over was banned. More generally, if someone complained about something – and really all of us had cause to do so – he met a noisy chorus of 'Tickets please' which shut him up at once, being the abbreviation of 'Give us your tough luck ticket and we'll punch a hole in it'.

Since our arrival in Belaria we found to our surprise, then increasing annoyance, that Art paid little attention to decent behaviour. He would lie on his bed for hours, not reading or talking to anyone and churlish in his replies. Or, out of the blue, he would say something disagreeable to one of us and seem to want a quarrel. Almost invariably he screwed up the 'stoogeing' arrangements when it came to his turn. This led to increasing bickering and resentment. Some of us, and I for one, started to push him back a little.

Then, in the turn from April to May, about three months after he had joined us, I woke up around two o'clock in the morning to find Art standing in his night clothes by a window he had opened with its shutter, despite the strict orders of the Germans. His face was bathed in moonlight and he was staring, tense and haggard, at the moon.

I stood still, watching him. Then he raised his head back and bayed at the moon, a long slow howl like a mourning dog, again and again.

He did not resist when we steered him through the door to the single room of the hut commander. The latter called the patrol guard, who sent for a medical orderly from the sick bay. As they walked with Art out of the hut, he saw the moon once more and began baying again. We listened to the slow, sad calling until it died away.

That day or the next, I walked round the perimeter to talk about Art with Doctor Montuiz, a major in the Royal Army Medical Corps, the head of the camp's hospital. He had been taken prisoner by the Germans while attending to our wounded soldiers in the withdrawal to Dunkirk in 1940, and served prisoners in various camps ever since. A couple of years earlier he had been offered repatriation to England but refused it in order to continue looking after us.

I told him about our dismay at Art's odious behaviour to us. We had answered in kind now and then. Could this have contributed to his breakdown?

Montuiz said that the percentage of young men in civil life in peace-time who break down like Art was statistically the same as in prison camp or anywhere else. Boats with weak planks go to the bottom and the others do not.

Art's disappearance was a relief for the rest of us. Being a prisoner at all was bad enough without Art on top; it had been hard to bear. Now a new man was posted to us by the hut commander so we were eight again and pre-Art peace prevailed.

It was hard to accept that confinement was certain, now we had been forbidden any chance of winning home around the magic corner. By May, some began to talk of escape again, but they were smartly overruled by 'China Bull'.

Probably he knew, although none of us did, that in the North Compound a 'revenge tunnel' was being dug, despite a certain amount of opposition, especially from the theatre boys under whose building it began. It grew slowly to a great dimension. But it was never used for escape: in the light of the advances of the Western Allies in the autumn any mass escape seemed to lose purpose, to say nothing of the memory of mass murder.

I believe that, in fact, the Senior British Officer of North Compound had no more intention of allowing his charges actually to use their tunnel than our 'China Bull' to build one, both of them for the same reason. He gave his a placebo and ours did not. We just had to grow accustomed to the meagreness, the dull background to our picture.

I and others began to find it difficult to look at the wire all around the perimeter. In fact an aversion to wire persisted with me for well over a year after the war, for example I would look away from wire running past fields seen from a railway train between Oxford and London.

What our Senior British Officer did do to help relieve tension was to persuade Belaria's *Kommandant* that walks in the country under parole should be organised. Only one German bearing a rifle would accompany us in case German civilians became aggressive.

Lots were drawn and I got a place on the second or third walk. That was lucky. My walk was in early July, 1944. By then, because of the ever increasing Allied bombing, the number of prisoners of war in Belaria had increased to at least 800 (by the time it was evacuated in January 1945, the volume had increased by half again), so one might have waited for the rest of the year for one's place in the walks.

When my turn came, it was a golden day, the sky was bright blue, leaves were shaking on the trees. It was a wonder to look across immense green fields with no wire to impede our vision. We saw women working in the fields, the first for most of us since our respective imprisonments began.

There must have been about fifty of us, walking in ranks of three for good order's sake, not marching but strolling with our German in the front. I was walking between two officers I did not know. The tall man on my left had bright red cheeks and wore, rather unusually, full RAF uniform, wings and a forage cap showing fair curly hair. We chatted for a while, then quietened to enjoy the walk.

I have a vivid memory of the officer on my right around an hour later. He was short, with bushy black hair, strongly built and staring at a young peasant girl as if he was mad. She was digging with a pitchfork on a patch of grass only a few feet from our road. Her dress was bright yellow, she wore a pink scarf round her head, she caught his eye and smiled and waved to him. She was certainly a foreign slave worker but cheerful for all that. He waved back, he seemed hypnotized by her, we passed her but he kept his eyes fixed on her until he could twist his neck no more.

At that very moment we heard a roaring motorbike, a German officer on it swept past us, turned and stopped our German guide, jumped to the ground and started shouting.

Our crocodile stopped. He strode back alongside it, loudly demanding in English, with a clear touch of panic, where was a name I could not make out. When he neared our rank, the well dressed Kriegie who had been walking on my left stepped out of the line and said he was the man. I then saw that the German was our Belaria Kommandant. With panic gone but rage instead, he shouted in German: 'Come over here or I'll shoot you at once. I'm not going to jail for a crazy young shit like you.' Our young friend strolled across and stood quite composed by him, smiling pink. The *Kommandant* bellowed for his trooper to come at the double. The trooper dashed up, stamped smartly, saluted and stood to attention. He was ordered to about face and march us all back to the camp at once.

As we left I saw our man standing there with the red-faced *Kommandant* glaring at him. The sun was still shining and so was our man.

When we got back to the camp, the miserable news was that our poor friend had left an envelope to be opened an hour after the promenade began. It said he could no longer bear being a prisoner, he was going to run away from the walking party so that the Germans would execute him for breaking parole.

As it was, the *Kommandant* and 'China Bull' simply agreed that he had a breakdown. He was not seen in the camp again and the word was he had been sent to join Art. What was sure was that the *Kommandant* put an end to the walks.

His anxiety had been well founded. To the satisfaction of the SS and the Gestapo, his previous superior, Colonel von Lindeiner, was by then awaiting court-martial, and probably a prison sentence, because of the mass escape from the North Compound. If our young friend had in fact made a bolt for it and got away from us, he would soon have been in the hands of the police. In that case he could well have been tried in a Nazi court outside the *Luftwaffe* for breaking parole and quite possibly executed, and the *Kommandant* would have followed Lindeiner into court martial, disgrace and serious punishment for permitting the walk and for negligence in allowing the escape.

In fact escape was at its nadir in the summer of 1944,* forbidden by the Senior Officers of the camp and faced with a strong probability of being murdered if caught.

However the mass escape of the North Compound had re-opened an old anxiety of Hitler's, that if prisoners, and especially officers of the Western Allies, escaped, they might promote sabotage and perhaps guerrilla warfare involving some of the millions of foreigners who were working against their will in the Reich.

Himmler, ever ready to expand his power, at once reminded him that he had long maintained that the camps of Allied officers – Air Force, Army and Navy – should be brought under the direct rule of the SS and Gestapo. Only their special methods could prevent the constant escapes which were a drag on German resources and could be a serious danger. Hitler agreed, and decided there should be an SS general in overall command of Germany's prisoners-of-war.

Waffen SS Lieutenant General Gottlob Berger was chosen. He disliked the appointment and complained about it to Hitler, who sharply told him to do as he was told. Berger hated the Russians but, in fact, went to great and successful

* On 20 July, 1944 Count Klaus von Stauffenberg tried to blow up Hitler, but failed.

lengths to negate a gruesome intention of Hitler and Himmler towards us. He began by shifting his new headquarters of prison camps to a place some miles away from his own Waffen SS military headquarters. And thereafter he ensured that the SS and Gestapo were kept well away from the camps, and the German military staffs were maintained over us as before.

Later in this story, he will be seen as the probable saviour of the lives of tens of thousands of Allied officers from one of the last orders of Hitler in the *Reichskanzlerei* in Berlin.

Soon after the aborted country walk, our clique broke up. In the Centre Compound most of our time had been spent together on hard escaping work, and in that half year, the second of 1943, we became a tight band of common purpose, almost a blood brotherhood, looking for the magic corner.

But in default of those activities, the bond had withered, all unnoticed, by the first half of 1944. Manser and I spent our time mostly reading and writing as if we were in college. Except for meals and occasional poker after lock-up for the night, we saw less and less of the others and only knew their interests were different from ours.

Increasingly, trivialities which would have passed unnoticed when we were exhausted from days of digging, grew into irritations, and some of those to anger. For example, Bert Larson thought it funny, after lights were out and he was in his upper bed, to drop first one of his big boots and, with varying timings, the other with solid thuds on our wooden floor. I, who slept below him, grew nightly infuriated by this. But there was no stopping him. Again, Hardy, the Texan volunteer in the RAF, had gradually turned against all things English and let this be well known. He once berated me for wearing our uniform Air Force tie. I said merrily that British officers always wore ties, he shouted in a rage that that was why they were hated all the world over. Manser and

I laughed and Manser grew sarcastic, an art of which he was a master.

Doubtless we irritated those two as much as they did us. Mo, good-hearted, placid Mo, kept out of it as best he could, and so did Field. But difference of ways of thought, of speech, of backgrounds, were beginning to scratch, and one morning in midsummer they all four packed their kits and moved into another mess without notice to Manser or me. We thought it was a little rough in view of all we had done together. But with hindsight it seemed just as well and our hut commander refilled our room with new airmen recently shot down.

Like most messes, we stuck a home made map of Europe on our wall and, with a bright red crayon, eagerly drew all military fronts to our advantage as they developed. When we arrived in Belaria we had only a serpentine line still far to the east of us in Russia. Then there were small red blobs in Normandy. Red bulges and arrows and spikes appeared until we had a hopeful hedgehog drawn half across France, whose bristles grew day by day and whose body fattened delightfully.

19

Rough Times

For a time there was hope that the Western Allies would crush the Germans before Christmas 1944. When the setbacks made it clear that this was not so and we knew that the Russian armies were advancing apace on broad fronts, we hoped they might be our liberators.

Their distant guns began to be heard by New Year's Day. Streams of German refugees from the east could now be seen on the lanes by our camp, horses or bullocks pulling carts with women and children and household goods, their men on foot or horseback alongside.

In the early evening of Saturday, 27 January, 1945, the German High Command ordered *Stalag Luft III* to get on the march immediately. In the North Compound the adjutant broke the news during the camp theatre's performance of *'The Wind and the Rain'*. The Compound began its move at about 1 am on 28 January, and by 4 am all were on the march – into the ice and snow.

Whilst the North Compound was being evacuated, before the Russians got there, a number of officers who had worked on their last and still unused tunnel obtained leave to hide in it to wait to be overrun. While the others were filling their rucksacks and sledges with all possible food from the Red Cross supplies, now opened for this purpose, they tucked themselves into the tunnel with plenty of food. However, when the SBO learned of it he ordered them to climb out and join the column.

The message got to Belaria later than to the other compounds. Now we, in our turn, began a frantic sorting of clothes and cherished objects and to calculate how much food one could shift on a home-made sledge. Finally, after many delays, Belaria was on the road in the evening of the 28th, heading into the bitter night of what is now once more called Poland.

A contingent of ill or, in some way, damaged officers remained in the sick quarters. This included me, making the most of my pelvis and back damages from the air crash in 1942 and hoping to stay until the Russians came and liberated us. It seemed a good, if rather unscrupulous, idea while it lasted. This was not long. In the first week of February we too had to trudge away from the camp, pulling kit bags on improvised sledges over the country lanes. Dr Montuiz, as head of the sick bay, acted as our Senior Officer during the journey. After some two hours, we came to a barely lit railway line running through trees in the country. There was a siding from the main track where we were packed into box cars which we were told were headed for Nuremberg.

Packed we were indeed. Whether from malice or imbecility, the guards herded so many of us into so few unlit cattle trucks that, when the doors were slammed and locked against us and our luggage, we were left standing tightly together. We could barely turn our shoulders and it was impossible to sit down. The train rode fast, bumping and swaying through a night of near torture for us, without water or sanitation. As it wore on several officers collapsed. They were held up by the body pressure of the rest of us as a slither to the ground was almost impossible and would have risked trampling. Vomit and involuntary defecation covered the floor.

When dawn came the train shunted into another siding among trees. The doors were opened and Dr Montuiz called

154

on us to descend and seek to defecate. I was standing near him when the Hauptmann in charge of the convoy walked up. In excellent German, Montuiz sailed in with a formal complaint about the scandalous conditions. He said that, if continued, they could well be lethal for some of us. With fury, he held the Hauptmann personally responsible. Unless the matter was remedied immediately he, Montuiz, would see to it at the end of the war that the Hauptmann would be tried as a war criminal by the British Army.

That did the trick. The Hauptmann had evidently been grossly negligent, so that the guards had been able to settle themselves in relative comfort in an ample number of wagons at our expense. There was no mistaking his embarrassment for what had happened, and fright for what might. In short order he had his men swill out the wagons we had been forced into. He added several more wagons for our use to the detriment of the guards and obtained a specific written pardon from Montuiz.

All I remember of the rest of the journey is one or two changes of trains on the way to Nuremberg and when we got there a march of three quarters of an hour to our new home.

We reached Nuremberg camp on 10 February, 1945, in the late afternoon, in drizzling rain and freezing cold.

The camp had originally been used for French officer prisoners. Since mid-1944 American flyers were being shunted into it, survivors of dead aircraft, as the US Air Force deluged their bombs in daytime from the German skies. And now the arrival of our own hundreds of more Kriegies, and similar marchers from the north and east, landed on the camp as cloudbursts. Bed bunks went from three to four above each other, palliasses were old and unsavoury, latrines were grossly insufficient, everything was dirty.

Also, rations became poorer than ever known before. Allied bombing on the German railway network had practically stopped the Red Cross deliveries which for years had

155

been the lifeline of our prisoners. The goon food we were given, never abundant, was still less now. Although the managers had to be prudent, eating into the Red Cross stores became inescapable in all the camps. Real starvation had not yet arrived, but it was growing thinkable.

We did not know it yet, but February was going to see the worst of our hunger: during that month, SHAEF* and the International Red Cross came to an agreement with the Germans for massive food supplies to be poured in by truck convoys from Switzerland, starting in the beginning of March.

Some ten days after our arrival in the camp we found ourselves in still greater danger. Under the Geneva Convention, prisoners of war were not to be held within three kilometres of major military targets. But our camp was well within the range of Nuremberg's great railway marshalling yards, about a kilometre and a half away. And, starting on about 20 February, the yards became the repeated target of RAF low level precision bomber attack by day, and heavier and higher by night.

Luckily for us, on the edge of the camp facing the yards there was a low hill which lifted and sank again like a sea roller. Inevitably, some bomb-aimers, our friends, made near misses from the target and now and then their bombs exploded the other side of the hill. Once, however, around midnight a real screamer seemed to be coming our way, and rose to a deafening howl. I was lying on my fourth level bunk, just under the wooden ceiling, and in a flash I was lying spread-eagled on the floor of the hut and never knew how I did it.

What dived was a shrieking plane, one of our own, and, seconds after, a massive bomb, missing the yards and

* Supreme Headquarter Allied Expeditionary Force

gouging a great quarry in our protective hill. A few yards difference and a lot of us would have 'bought it', as we used to say. Sooner or later a bomb seemed certain to arrive a little higher, and over the hill and onto us.

A different memory remains of standing outside our hut on a fading pink and soft blue evening, and watching the high level bombing of Nuremberg itself, the city just below the horizon. As twilight came, the scene opened with a few small lights in the sky, floating down – the marker flares. At once searchlights lit, interweaving and swaying above, and hundreds of bright needles in the sky, the endless flak shells bursting at our planes, too far from us to hear. As the evening deepened red stabs appeared on the ground horizon, here and there, slowly growing from small to huge, and weaving an orange carpet for the city to burn its enormous fire as the night drew on. Throughout the drama, now one, now two or three at a time, small red embers floated down, turning to brighter and longer brush strokes – our bombers crumbling as they and most of their crews were lost. From this time until our leaving the camp, we had to endure several near misses at night from 'friendly fire', but none as bad as the crashed aircraft.

*   *   *

Since the tragic speech of 'China Bull' in 1944, we had obeyed his order to cut out escaping. After the vacuum of summer and autumn, we had hoped that the advancing Russian armies would liberate us in January. But we had been force marched west, away from that, in the bitter winter.

It would have been easy to drop out from the march. But our Senior Officers again ordered us not to do so, in view of the still long distance from the Allies, the danger of freezing to death alone when moving cross country, and the reasonable chance of being strung up by the SS if they got their hands on us away from the columns. We complied.

But now, in February, 'China Bull' was away with the main

body of Belaria in the north, the Allies were much nearer, our 'friendly fire' was only too frequent, and the camp was dreary beyond belief. All this pointed to getting moving again on the escape front.

And possibility came quickly. Two French officers told me that some while ago they had helped a couple of their friends get away through a manhole into a drain pipe under the camp and beyond it, giving into a larger sewer system in the country. A few other manholes marked the line of the pipe until near the edge of the camp.

They said the drain pipe just allowed a man to get his bare body mobile in it. It was empty but mechanically flushed with water at hourly intervals. Anyone inside the pipe then would be drowned. The trick was to wriggle from one manhole to the next and stand up in it to let the flush go by, and then to wriggle to the next and stand up and wait again, and so on under the camp and on to the sewer.

We walked to a metal manhole mostly covered with grey dirt, not far from wires which separated us from another compound for Russian prisoners whose lot was even worse than ours. Then we walked some seventy yards to the next manhole parallel to the wire.

The Frenchman said there were three or four more to the end of the camp. Their friends had entered through the first manhole we had seen because it was near some of our huts, while the other manholes were in open ground, much exposed to the guards.

They left me then, wishing me the best of luck. They said they did not want to be involved in this for a second time because their families were in France and under daily exposure to the Germans.

I talked about this to Guy, an older officer, whom I had known in Belaria and suffered with in the hideous train to Nuremberg. He said he didn't like the idea much and certainly wouldn't go along with it. But he wouldn't mind helping me lift the manhole cover when the goons were not around.

Before and after this period, Guy became an unforgettable friend of mine, giving me advice under many headings for after the war, some of which I sorely neglected. After Eton and Christ Church, he had become a member of a great brokerage firm in the City until the war. Then came the Air Force in which he was a pilot and earned the DFC and bar before he was shot down. When the war was over, he rejoined his brokerage firm. I give no further name because, a few years later, Guy killed himself, to the grief of his many friends.

Next day, with a couple of other officers for cover, Guy and I raised the heavy top of the chosen manhole with a wrench. Hanging my boots under the lid and armed with a torch and a towel I called them up to close, but please hang around. I had a hammer to bang up if need be.

I was standing in the torch light on a cold stone floor, upright in a stocky metal cylinder. On my left a wide wet mouth of orange pipe yawned at me, and its twin on my right the same, in the line of the camp and, hopefully, our exit.

I stripped bare and entered the pipe. I could just move without my shoulders touching its sides. But I could not bend my legs on all fours so had to creep with them half bent, and pulling by the arms.

That crucial information obtained, I shoved myself in reverse to the manhole and upright in the pipe to wait for the onrush of water I had heard about. When it came it was strong up to my knees but not enough to sweep my legs from under me. To my considerable relief, with memory of old times, all the water that came seemed like dishwater, and none of it was shit.

Next morning, thanks to the two helpers (Guy had dropped out), I went down again to count the intervals between the gushes. Three tedious hours later showed that it was indeed an hourly clockwork with about six minutes for the gush to clear. I though a man would have to pull hard and steady to be in the next manhole before the gush to survive. In the manholes he would need to stand for six icy minutes all naked, and be back again without loss of time.

My helpers saw me up for lunch and a rest, and then dropped me down again to discover what thirty minutes of exploration would be like. When the next gush was spent, I half crept along. There was no bad smell. But my neck was aching when fifteen minutes were up. I went backwards by shoving my hands in reverse which took more like twenty than fifteen minutes. It had been no picnic.

The two helpers were really good natured in all this. I have forgotten the name of one but the other was a young English pilot, Matthew, medium height, blond with a boxer's broken nose.

If the French officers could do it, it could be done again. But I thought it would be hellish rough. I was in my bunk, and 'lights out', and then the idea came: roller skates could be the name of the game. Buy them from a corrupted goon, fix the skates to belly-sized planks and the planks firmly to our bodies then, hopefully, pulling ourselves along should not be too difficult. Find the right goon I thought, find the right goon, and muddled into sleep.

Awake next day, things did not seem so easy. To corrupt goons in the other camps had taken plenty of time. During the morning I was able to chat separately with a guard and a workman. They were quite friendly and happy to smoke a cigarette with me, but I would clearly have to find others to get anywhere.

Meanwhile we had to progress underground. The day before it had been just endurable to slither forwards. It had been pulling one's self in reverse which had been well nigh intolerable.

Accordingly Matthew, the other helper and I went back to

our water hole, checked that the dust on the top was unchanged and walked to the next manhole along the way out. It was a sunny day and we sat on the sand by its cover, clearly outlined in the dust. There was no one about. We quickly removed the dust and used a spanner to see if the cover was locked. It was not. We lifted one side of the top for an inch to be sure, put it back, smeared the dust over it and walked back to yesterday's hole.

I intended now to go underground the whole way from the first to the second manhole, stand up in it, turn round, and creep back to the first. The pipe was feasible but only head first. With that qualification, we had a more than usually tight, ready-made tunnel. We could not understand why it was only the two French officers who had found and used it. When the time came we too should certainly be able to get to the sewer.

Next day Matthew and the other man lifted the top, in I went, down went the top, and, practised as we had become, the whole thing took barely one minute. I did my solitary striptease again, waited for the hourly gush, and then forced my way along the pipe to underneath the top we had examined the day before. Matthew had meanwhile strolled back and sat on the ground smoking and reading a book. When I reached his hole from underneath my arms were still working and it was less than forty minutes. I hit the spanner up, Matthew hit his down, and both of us were relieved. The next gush came as usual, and I had not much trouble on the return. One could perhaps forget about the skates.

Then Matthew said he would like to come too. I had hoped he would but didn't think I should push him. Now I was delighted. Crawling all alone along the pipe and thereafter in the sewer would have been pretty disagreeable.

Escape called for preparation without delay. There was no Escape Committee as before, no obliging seamsters, map makers, or any food store upon which to draw. We would have to make light bags to contain a blanket each, a food pack, gathered somehow and a water bottle and shove them ahead of us down the pipe. We would need well-constructed miners' torches.

When we got away from the sewer, we would move cross-country towards the advancing Americans. According to our squeaker, they were moving east in the direction of Nuremberg with all majestic speed.

So far the plot has been remembered clear and bit by bit. But it is now long ago, and time has mercifully forgotten most of dreary Nuremberg in detail.

At some point a Polish officer sold me two thick slabs of American Red Cross chocolate for twenty-five pounds, more than a week's pay in those days. I gave him a cheque on a scruffy piece of paper, addressed to Lloyds Bank, Oxford. In the autumn of that same year, when I was up at Magdalen again, my bank manager rang me to ask about a weird paper demand just received for £25, and what should he do? The answer of course was he should pay it, and please let me have the address. Two weeks later, the Polish officer (his name is now lost for ever) came to see Mary and me for a few days. He was wondering whether to stay in England or return home. I advised him to stay, for communism was growing fast in Eastern Europe. As I heard no more from him, I could only suppose he had returned – a sad feeling.

In the camp, a day or two after my purchase, I looked in the box I had by my bunk and found that the chocolate had disappeared. A theft of that sort was serious. The thief was somehow discovered to be a French officer and I remember being at a brief parade when he had to stand to attention while a senior French officer read out that he would be court-martialled when the war was done. The shame and horror which smeared the thief's face was appalling. I hoped he was insane, and that he would be so regarded when the time came.

I have no picture of how our food and kit was eventually

put together, but I know it was, and looked after by a squadron leader for us. Someone else was making the bags which were to be pushed ahead by our hands, push by push, until our pipe opened onto the sewer.

Then, beyond belief except that it is true, I came across a cheerful goon guard walking around, it seemed idly. We started to chatter and smoke my cigarettes and after a few minutes, all or nothing, I said I would give him several boxes of cigarettes in return for two sets of old roller skates if he could find them. He laughed like hell, and 'For God's sake how will you skate in all this muck' pointing at the dust, and, yes, he did have some skates among other junk in a shed in the backyard of the house he got home to most nights. He would be happy to oblige for a decent lot of cigarettes, and would I please also give him some pieces of soap.

From the endless Red Cross provender which had to be abandoned on leaving Belaria every one of us had packed as many cans of cigarettes and stocks of soap as we could jam into our carry bags, both for our own needs and for barter, so I also agreed to the soap, and would he be quick about it? The answer was yes, as soon as he returned from a three day leave he had from that evening which he was going to spend with his family.

If all our kit had been ready, we would have left without further ado. But I reckoned we would still be waiting when he got back. The skates might be easy to fit, and anything that eased the way to what was going to be extremely demanding was a friend of ours.

20

A Long Walk

Two days after the goon's leave began, the camp *Kommandant* told the Senior Officers that we were to evacuate the camp the next morning, 4 April, at daybreak. This was a much better way out than worming our way down the long, dangerous sewer. All the Kriegies spent hours that night turning sheets into carry bags for their precious kits and the Red Cross parcels, now opened to the camp stores, which would provide the food we would need on the journey.

I remember packing a pair of shoes in case my boots gave up, the notebook of my poems and the Oxford English Book of Verse, my beloved cashmere dressing gown along with a large amount of cigarettes and soap packets for barter. Climbing up to my fourth layer bunk for what was left of the night, I closed my eyes, much cheered to be marching out of the camp, never mind the guards, we would see about them later.

Mercifully the Royal Air Force did not bomb the marshalling yards that night, with yet another chance of a near miss on their target and the possibility this time of it landing on us.

After much confusion and endless standing about next morning, at eleven o'clock on 4 April we stepped out of the dreary dust which covered the whole camp, and into a side road stretching straight before us under a pale grey sky. The sun was hidden, so the colours of early spring were subdued. Nonetheless, trees were soon standing on the side of the road and their trunks and branches were a sweet, clean brown, the young leaves were bright green and yellow, the shining, growing grass looked juicy to us who had scarcely seen vegetables in the last two months, and above all there was no wire. The march was in separate columns, each a few hundred officers or other ranks, so that the leading columns, among which we found ourselves, were miles ahead of the last. Our column consisted of those Royal Air Force officers who had been moved from Belaria to Nuremberg, and large contingents of American and French prisoners who had been there when we arrived. In the columns, most Kriegies grouped in the same messes as in the camp for friendship, mutual help, and efficiency of makeshift cooking.

We had a guard walking alongside us every forty metres or so. Some of them were disagreeable, shouting at us to hurry, smacking the butts of their rifles and warning us in pidgin German how they would shoot us if we tried to run away. We kept quiet. Soon our road was running through woodland, and in the first hour of our walk I saw a bomber pilot whom I knew slightly, a tall, athletic English officer, slip calmly into the trees on a bend when our guard was looking the other way. Later, when everything was over and I was being de-briefed in England, I was delighted to hear he had made it to, and somehow through, the advancing Allied lines. That was good work indeed for a pedestrian across highly hostile country, many miles from where he started.

But our general view was that it was wisest to stay with the column. We had heard that detachments of SS were roaming about, especially in forests, to catch and hang German deserters, and they would certainly do the same to us if we were caught. The word had quickly spread that our destination was Moosburg, near Munich, and if we stuck together we had a fair chance of getting there intact.

We plodded through the first day with irregular short stops to rest and, once in the afternoon, for a halt of some fifty minutes. There we drank at a group of houses with running taps, refilled the water carriers which each of us had, and sat down to eat what picnics we could produce from our bags. By the end of the afternoon we were pretty tired, having marched well over twenty-five kilometres for the first time in years. Some of the guards, many of whom were twice our ages or more, were in an even worse plight.

It was also noticeable that the bravos of the start of the march had now stopped throwing their weight about and were plodding silently and evidently downcast. Away from their camp at last, the penny seemed to have dropped on even the stupidest of them that the Russians were moving relentlessly westward, and the British and American armies eastward into Germany. A few more weeks and the Reich would be crushed in the continental nutcracker. Guards who pushed us around now would not have shining prospects later.

The first night of the march saw us billeted near a village, in huge farm barns of stone and wood. We wrapped ourselves in the blanket which each of us had taken care to bring, lying on the floor – some on straw and some without – all dead tired and mostly dropped off to sleep.

We were up early the next morning. The sun was shining, and in my head it remains so every day thereafter through our journey. At the beginning of the morning, some of the younger guards sought to chivvy us along, but all vestige of threats had ceased, they stopped soon and our column settled to a speed, still somewhat above a ramble but never more than a comfortable walk in the country.

At about ten o'clock, up in the sky above where we had started yesterday, we saw a huge raid of US bombers pouring their bomb loads on to Nuremberg. Later in the afternoon, the tragic word came to us that, in addition to the city and

166

the marshalling yards, our camp had at last been badly hit in the melée, with many US prisoners and many more Russians killed.

The march began to straggle, slowly regaining cohesion when those in front had been halted for the night. This concertina act was repeated every day until the end at Moosburg.

I found I had over-packed and Guy insisted that I should purge some weight. I can still see my Oxford Book of Verse which he had put forlorn in wet grass on the side of the road.

The second night was spent near a fair-sized village. There were several big barns and some smaller huts which the guards commandeered for themselves and their charges. Our group, however, found a small Baroque church, dimly lit by an electric lamp. We lay in our blankets by its altar, well worked in pink and white marble with charming gold decoration. We relaxed in the flickering shadows and candle lights, talked a little and turned to quiet sleep.

Around mid-day next morning, a friend of mine and I saw a light four-wheeled wooden cart outside the fence of a large villa by our road. We at once commandeered it, each grabbing a shaft, and overtaking some of our column as fast as we could without trotting, until we were well up the road away from the owner. We then dawdled until the rest of our mess caught up with us, all our luggage was bundled into it, and we were very popular: eight Kriegies could now walk with nothing on their backs in return for easy pulling in hourly turns. It made an enormous difference.

By now the guards were altogether more relaxed, seeing that no one was escaping. When we passed through villages we bartered in the bakers' shops. For years on both sides of the war, food had been seriously less than in peace. By this time it was even scarcer, or worse than scarce, in much of Europe. But, except for the actual battle areas, farmers throughout Europe always built secret food reserves for home. As did most others in those days, they liked to smoke. But tobacco was hard to come by in the Reich by then so our cigarettes were our gold, as were pieces of soap or slabs of Red Cross chocolate. I remember buying loaves from a baker woman in return for cigarettes, and when a lady shopper saw it, she insisted on giving me a rather good big cake for more cigarettes. Frequently, we wandered from the column to farm houses across the fields for eggs, barley, and all kinds of vegetables.

Nearly all the civilians we met were polite and many were talkative and friendly. A few slammed their doors on us. And once two from our mess and I knocked on a comfortable stone house to barter as usual. A middle aged woman opened the door, her face went aghast and in a hissing, desperate whisper said we must go immediately, 'The Nazis are inside, they will shoot you if they see you', and we backed off fast. We were not home yet.

At the end of one afternoon, some of us settled in a field below the garden of a farm house. So I went to the door and bartered cigarettes for bread and dried fruit. Talking with the young mistress of the house after the bargain – her husband was away in the war – she and her sister poured out the woes of their heavy life, their longing for the war to stop. They asked if I had photographs of home. I showed them one of Mary and myself cutting our wedding cake with her father in their garden, she in a white wedding dress, I in my cadet's uniform, my mother, Tony Garton, my best man, also from Magdalen, now a Grenadier, young people all around, a happy English wedding party. I heard her say in German to her sister 'Not like us beasts of burden!'

Later, near evening, I walked from their house down a field to a cattle fence where I saw a fair-haired young man watching me over it. As in most farms, when the men of the family were away in the war, he was a forced labourer from Poland. He was poorly dressed and very muscular. He spoke well of the sisters, who gave him adequate food in return for good work. But he had bad tales to tell of some other German farmers, and especially of men who were still working their own farms, mean with food and ready with whips for workers like himself, and even worse with young Polish women, totally enslaved. He said that he and his friends would 'teach them something' the moment the war was over.

Another night, we were placed in a number of smaller houses as well as barns. Four of our mess were put into the dark house of a large grim peasant, straight out of Pieter Brueghel. We sat with him, we ate his food, he had huge red hands, we felt he did not like us, his conversation was careful, he seemed to be wondering what the future might bring. I discovered why when I entered a cow shed lit by one small candle, and found a couple of Polish forced workers, a man and a woman with poor, sad faces who told me the farmer was a very cruel man, fond of the whip, and they would see to him at the end of the war.

Indeed, in the first days of German defeat many slave workers in the deep countryside 'saw to' unkind masters, some frightfully, pinning them to barn doors with pitchforks through their chests.

One morning, a German sergeant from the security force, last seen by us in Belaria, somehow fell in with me. He was well built, self assured, about twenty years older than me and wondering what the future would bring. His main concern was his wife who was walking with him, also in military uniform, a pretty woman about my own age. I remember the three of us talked a great deal together. Once, perhaps to show off to me, he said that when the war was lost he might join the Werewolves in the forests, reputed guerrillas said to operate behind the British and American lines. She turned on him at once, laughing, and said she would never let him go 'mein Schatz' (my treasure) and he had to look after her. A weird conversation set in and my last memory of them was her half smiling, half weeping, saying 'if only the three of us could fly away to a desert island and be happy together', an observation to which I found it hard to find an answer.

A few miles further on that day, as we trudged by green fields and sprouting trees, what was nearly a peaceful stroll along a country road was suddenly torn by shouts to drop into the grass, two US fighter planes had just buzzed the column behind ours, firing their guns, they were turning and coming again. We flattened in a flash and down the planes came indeed, but this time waving their wings at us, roaring over the field twenty feet above our heads. Later we heard that they had killed two officers of the Royal Navy who were away from the other column and who they had thought were Germans.

It sounds amazing, but later still that day, some of the older guards were so tired that we allowed them to put their rifles and other chattels on our cart until we packed in for the night. More, for a tiresome hour we let one guard, who was in really bad shape, sit in the cart and be pulled by us until we could dump him on his sergeant.

What turned out to be my last night on the march began like any other. Our mess moved into a school room to sleep on the floor after a brew of last year's barley, lifted from a farm we had walked by and cooked on a camp fire in the school garden. A few hours later – though it happened to none of our other diners – I woke and felt so sick I could barely jump through the door into some bushes, where barley and everything else came up again and again. I then collapsed in all of it until first sunlight and staggered to where I somehow knew the *Luftwaffe* Hauptmann responsible for our column was sleeping. I told him I could not walk today. He took one look and said all right, he would fix it, and go back to the room I had slept in. In half an hour a *Luftwaffe* goon came to me, picked up my kit and led me a few steps into a small, open military truck.

No doubt it had other purposes as well, but he drove a good many miles along the track the column would be taking, turned into a large farm and took me to the main door of a massive old stone house. He spoke for a couple of minutes to a middle-aged woman who was evidently running the place and drove away. After my night exercise, I was of course inelegant from every point of view. She beckoned me into the hall, laughed cheerfully at me, unhooked a man's loden coat and told me to take off my boots and all of my clothes, leave them on the floor, put on the loden, and follow her into the garden.

The sun was well up. We sat in comfortable wicker armchairs near an empty stone trough and she called a forced labour girl – quite pretty, about twenty-two, Yugoslav, dressed in decent red and brown house clothes, an evident slave, but a well treated one. My hostess told her to get all my clothes washed and send two other girls to fill the trough with hot water.

I spoke my thanks to her – by now my stomach had come to rights – she was pleased I knew German, and even more pleased when I gave her a tin of English cigarettes. While the girls were walking to and fro with steaming buckets she embarked on cheerful chat with me about how marvellous it was that the war would soon be over, my mother would be very pleased to see me, what was I going to do after the war? She told me the driver had said I was an English air terrorist, I didn't look like one, and so on, until the trough was full. Then she took my elbow, pulled off my coat and watched me with close interest clamber and sit down in the bath water when she handed me a piece of soap – very scarce in those times – and watched me using it as if I was in the nursery.

The girls brought me towels, how nice to feel clean again. Then back to the coat and the chairs, the sun was hot, we chattered endlessly, ate sandwiches and even drank beer – what a joy after so long. At about mid-day, the Yugoslav girl brought back my clothes, all clean and fairly dry. I dressed beside the trough and then she brought me a bowl of soup. My, by now, friend looked at her watch and went into the house for a few minutes. She came back and said another truck would be collecting me soon. Sighing she said she wished I could stay with her – and so did I by then. The truck came, I went, she waved.

21

Out At Last

This time it was a medium-sized truck with half a dozen goons in the back who were being moved in advance of the column to our objective camp, Moosburg, now some twenty kilometres away. My *Luftwaffe* Captain had told them to take me along.

When I got there, ten minutes of being locked into the camp was enough to see, hear and smell how unpleasant it was, compared even with the Nuremberg camp we had left behind us, and all the worse for the quasi-holiday of our walk to it.

Since December, 1944, prisoners-of-war in their thousands had been marched from the eastern Reich, away from where the Russian armies were advancing, into north, middle and south-west camps. And these camps had already been overcrowded by the major intakes of prisoners-of-war during 1944. Consequently, by now they had been swamped. Sanitation, by the end of March insufficient, was now overwhelmed, water was only obtained by long queues from too few taps, food supplies inadequate and, indoors, lice a-plenty.

Moosburg had all of that and, in addition, sleeping quarters so wildly insufficient that swarms of us spent the nights on the ground outside the huts under makeshift canopies. On the other hand, to some of us at least, it occurred to be thankful that, compared to Nuremberg, we also had no friendly British or American bombers likely to blow us to pieces by near misses.

Two days later, I was greatly cheered by suddenly meeting Bill Manser out of the blue. He had arrived in the camp that afternoon and was looking for me after learning from the SBO's office that I was there. It was a joy to see him, last seen leaving Belaria at night in January, for forced march and train to Lückenwalde, *Stalag Luft III-A*, a hundred and fifty freezing kilometres away.

He told me that a few days ago the German High Command had issued an order that the two joint Senior British Officers of Lückenwalde, Group Captains Kellett and Willetts, were to be transferred immediately to Moosburg. Their escort was a *Luftwaffe* Major with a sergeant and a corporal. Manser was assigned to them as an interpreter and they had an obsequious British batman for their menial needs.

The party had travelled by train as best they could, given the multiple destructions which the Allied Air Forces had inflicted on the German railway network. To Berlin first where they walked through devastation to a different station and then on through a maze of other lines, picking their way round bomb damage to Czechoslovakia, then a great meander across south west Austria changing at Linz and Salzburg, detraining at last in Munich, and into cars to the camp.

At the end of this quite dangerous odyssey, the two Senior Officers, to say nothing of Manser, still had no firm idea of why they had been despatched on it. One guess was that Manser's presence with them might imply negotiations needing translation with senior German officers as the end of war approached. But then there were other British and American senior officers already in place and other interpreters available if required.

Manser told me of the tough time the thousand RAF prisoners from Belaria had on their winter way to Lückenwalde, and their near starvation when they got there, along with some 15,000 other prisoners of many nations. Red Cross parcels were failing and German rations scant.

I countered with our spring saunter and the pleasures of barter. Then the talk turned to escape. On the one hand the Americans were advancing to the north of the Danube. Their general progress would evidently relieve us soon. All we had to do was wait. On the other hand the conditions of the camp were foul and, after all our efforts over the last two years, it would be agreeable indeed to make an escape while there was still a war to do it in. It would be escape for escape's sake, turning the magic corner at last.

When we turned in that evening, on the ground out of doors under a canopy, we had pretty well decided to have another go. We awoke next morning into a certitude: rumour was raging through the camp that we were all to be marched away again. Much worse, the SS might be planning a last stand in the Bavarian Alps, the Redoubt, and for it would gather all the British and American officers in the Munich region as hostages. If there was not time enough for all, then 'British flyers' would be specifically required, and it was clear who that would be.

Our officer compound, despite the gross overcrowding and unusual squalor, was as closely guarded around its perimeter as the others we had been in – warning area, high double fences of barbed wire, sentry boxes with machine guns – the lot. Our compound adjoined another for British and French NCOs, also closely guarded.

However, close to our compound there were less guarded settlements for British and French troops, along with other nationalities. Every day they went in working parties into the countryside and local towns and came back to sleep. Some were also sent into our camp for humble tasks like cleaning the latrines as well they could in all our chaos and making some repairs to the wooden huts, which were fast collapsing from the overcrowding.

Two British sergeants were in charge of these particular work parties which were mostly made up of British and French troops. One of them told Manser that he had it from a French prisoner that there was a German NCO who hated the Nazis and, if a couple of British officers could get out of the camp, he would be able to get them into a safe refuge for reasons not yet known.

This offer sounded by no means watertight. The mere possibility of the SS and the Bavarian Alps was a shuddering picture. Once out, we could distance ourselves from the German if we thought the game was fishy. So we accepted and thanked him for the offer. It all went smoothly. The next day the Sergeant brought in a party of French prisoners who were recorded at the gate between the two com-pounds. When their time to return came Manser and I, wearing French uniform jackets, simply took the places of two French prisoners who were happy to stay in the officers' camp.

We spent the night in one of the sergeants' huts. On entering it the smell of their Gauloise cigarettes brought waves of nostalgia for pre-war schoolboy holidays in Paris. They shared their meal with us hospitably. Then a sturdy, dark-haired French corporal invited us to another room like a small office. He left us and a few minutes later returned with a tall and rather portly, grey German corporal.

We were to hear much of this man's life story and inten-tions before long. At this stage he limited himself to saying that if we were ready to join him his English friend – our Sergeant – would take us out of the camp tomorrow in a working party of French troops to unload goods from a freight train half an hour's march away. Just before where the train was standing there was a large wooden warehouse. The

Sergeant would point it to us. We should slip into it and await nightfall when he would come for us.

Next morning, a pure, sunny day with a blue and golden sky, Manser and I marched along with a Frenchman, three abreast, towards the gate, we looking exactly like them in our workmanlike uniforms. An Unter-Offizier was standing on the left side of the gate. He checked the paper the Sergeant handed him, gave it back, and the men ahead of us began to march through.

As we drew nearer my heart dropped. It seemed for a crushing moment that 'Kriegie luck' might stop us yet again. The man at the gate was one who had good reason to remember me. During the march to Moosburg the day before I finished the journey in the lorry, I heard him, unlike most of his fellows by then, berating one of our officers who was suffering fatigue. In so doing the Hun used 'Du' to him.

The officer didn't care, even if he caught the language. But on the principle of always being as unpleasant as was safe to a German who was being unpleasant to us, I berated this goon well and truly in the harsh tones of an angry Prussian officer – which I was rather good at by then – for breaking what he well knew was their own rule to use 'Sie' when addressing a British officer. He had looked as if he would like to kill me, but had said nothing. If he now caught sight of my face the game would be up.

So with Manser on my left and a Frenchman on my right, I continued at our smart pace, turning my head slightly away from the guard avoiding eye contact, but careful not to overdo it. As we neared him, I pulled a handkerchief from my pocket and blew my nose, covering much of my face and, trumpeting vigorously, marched through the gate.

For a few seconds the skin of my back seemed to shrink before a shout that did not come, and then the iron gate clanged us out, after near two years of in. For a strange and lovely moment my walking felt like marking time, the gate behind us moving back, the sentries turned to dwarves, the guns to toys, the camp to farmland. Then 'click' we eyed each other, smiled and went, swinging our legs, into a new chapter.

22

The End Is Nigh

As we neared the railway, our Sergeant led the column along-side the warehouse and we slipped away behind it. The rest of the day was spent in the recesses of the huge, and apparently unused, wooden building hoping that our German friend would turn up when the time came. (Manser and I called him Jumbo between us.)

Slightly before dusk, he arrived in his corporal's uniform, carrying a rifle. He showed us an official form ordering him to conduct two French prisoners-of-war to a particular place where they were to work. He said he had forged the camp *Kommandant*'s signature and he made us repeat our respective French names several times in case we were halted. But we were not, and soon arrived at the house where we were to be hidden. It was on the edge of a village some three hundred yards behind a main road, a segment of which could be seen from the back yard. Our host was introduced as Willy, and he and Jumbo led us into Mrs Willy's kitchen.

It was of modest size and sparkling clean, and she a peasant woman with smooth, peachpink cheeks, twinkling blue eyes and a beaming smile. Their two early teenaged sons were watching us in awe and admiration. She kept chickens and we sat down to a fine supper of fried eggs and toast with a slice of ham each. These were the first eggs that Manser had eaten since his capture two years ago and, though I had bartered several during my recent promenade, they were still a treat.

During the evening we learned that Willy was a machine worker and that before he had inherited this house it had been a small garage, until the main road was built, which took all the business away. Before Hitler came to power Willy had been a moderately enthusiastic trade unionist. A couple of years of Hitler, and he was sent to a concentration camp because he had drunk too much beer one night and cursed the Nazis in a pub. He was released after a year and his general dislike of National Socialism had crystallized into violent hatred. He had been working with Jumbo in some way for the last year.

He had also dug a safety tunnel in his back yard, carefully and slowly, using his sons as sentinels. Over the tunnel mouth was an old car from which all the wheels were removed and the floor remained. A trap door was in the floor, lifted by pressing a particular part of the window frame. Under this was another trap with earth glued on it. On undoing a hidden bolt it opened down to a short laddered shaft which gave into an underground room.

He told us that he intended to hide in it if he was summoned to one of the Home Guard units into which old or unfit men were being driven during the final months of Germany's defeat. Luckily for him, any resistance activities he was involved in had not been discovered so far and we were to be its first occupants.

We ate and smoked and laughed in the kitchen and felt happy. Then Jumbo departed, saying that he would soon visit us again to have a talk, and we followed Willy across the dark yard, down through the floor of the car and the trap of the shaft and into the cabin he had described. It contained two small beds, blankets, a latrine bucket, one electric bulb which Willy had tapped from the cable underground to his home, and little else. The walls were plated with metal sheets and from the moment we occupied the cabin we lay in icy dampness from them.

After the first night, Willy brought in more blankets and other bed clothes for us to wrap against the damp, which of course was a new discovery for them as no one had slept there before. From then on, before dawn and after dark he or his sons used to bring us food and the latest BBC news, and empty our latrine. For a couple of days we read German books, talked endlessly to each other for want of anything else to do, and slept as much as we could. But throughout our stay the sound of dripping water punctuated our lives like the ticking of a clock.

Jumbo then visited us, climbing down the shaft to sit on one of our beds and explain his situation. In civil life he had been a publisher, and when the war was finished he would tell us his name and more about his life and family. Since Hitler came to power in 1933 he and all his best friends had detested the Nazis and their hideous policies, had talked about it in secret meetings, listened to the BBC and kept away from the regime as much as possible. He had managed to avoid serious suspicion although some people shunned him. He was conscripted when the war came, but his poor health kept him from the fronts and he had a number of lowly clerical posts. Two years ago he had been posted to Moosburg near the city of Munich where he and his family had always lived, and there appointed as an insignificant guard, more clerical than active.

He said that during that time he had built the nucleus of a small freedom party, determined to put their democratic ideas into political practice the instant the war stopped. He said that the assistance he was giving us would help establish his credentials when the fighting ended, and help him and his group to be disentangled from the inevitable interrogations of the occupying power. This seemed all right for us, though a shade less glamorous than the original tone we had heard in the camp. We little knew.

181

The next week or so went by slowly for us. But meanwhile, as we learned later, affairs in the camp had been verging on the desperate. A day or two after we had left, the Camp *Kommandant* received an order from on high that all British, American and French officer prisoners under his command should be ready to march south at short notice. Specifically included were all those of *Stalag Luft III* who had come to him via Nuremberg.

He at once summoned the Senior British and American Officers and relayed the order. He said he would inform them of the destination as soon as he had been informed himself.

The Senior Officers were dismayed. There was no known Kriegie camp south of Moosburg so it seemed only too likely that the *Kommandant* had the rumoured Redoubt in mind. It was clear that a refusal to move would be met by armed attack in the camp. So the orders to get ready were given.

This explained the apparently meaningless journey of Manser's two Senior Officers from Lückenwalde to Moosburg. It would be impossible for many of the northern mass of Kriegies to be joined to the southern group in time, but their two senior heads could be dropped into the plan like pinches of salt, to say nothing of Manser's.

The Senior Officers were probably developing decisions for mass breakaway before getting to the destination unless they were heavily surrounded all the way by SS, and individual groups of Kriegies were certainly doing likewise. After another anguishing twenty-four hours, the order to be ready to march from Moosburg was countermanded.

Our colleagues' anxiety had been well founded. In the Nuremberg Trials of 1945 and 1946 it was stated that in the time of Hitler's agony before his suicide, he broke down and charged Goering, away from Berlin, to sue for peace from the Allies. He also charged Waffen SS Lieutenant General

Gottlob Berger, master of all prisoners-of-war, to take a large number of officer Kriegies from southern Germany into the Bavarian Alps as hostages in the Redoubt. A truce was to be demanded from the Allies, but not the Russians. If it was not forthcoming, they were all to be killed. The target figure was some thirty-five thousand prisoners.

At the Nuremberg War Crime Trials, Berger was sentenced to twenty-five years imprisonment. Two years later, he was released when his true behaviour, towards the Western prisoners-of-war at least, was recognized.

From the moment Hitler gave him the iniquitous order, Berger was determined not to execute it. After the War Trials, it emerged that Eva Braun, Hitler's mistress for some ten years and wife the evening before their joint suicide, had heard of his plan. Horrified by it, she somehow learned of Berger's decision. They talked and decided that it was essential to avoid this order falling to some other high-ranking SS officer, who might actually get it carried out.

Accordingly, she had fixed a personal meeting for Berger with Hitler about details of the matter. During their talk she brought a typed command into the room concerning the order for Hitler to assign it further to Berger. Hitler signed it and, the meeting over, Berger turned on his heel and left the room and Berlin. By the time of the planned march to the Redoubt in April, he had the advantage of being in the south, some seven hundred kilometres from Hitler, and the lives of thirty-five thousand officers, although they little knew it, were safe.

As for us, we damply rolled along another day or two, sustained by Willy and his boys, and rather bored until one evening came when Jumbo visited us in great excitement. He was in his uniform with his rifle, all of which he took off for us to keep and donned civilian clothes from a small rucksack. He said there was going to be an uprising in Munich next day, ahead of the arrival of the Americans, and his group was to participate in it. He left his kit with us and went away.

Later next day, he returned, bringing a small German with him. When he stepped down the ladder and stood before us his body was literally shaking from shock and fear. The rebels had taken one of the main broadcasting stations in Munich, proclaiming the revolution of a democratic Catholic, Bavarian state, and calling on the Americans to hasten to reach them. But Nazi forces counter-attacked and killed all twenty of the insurgents. Jumbo, who had been near but not on the scene, was now hurrying to get back into uniform and the camp for fear of being missed, suspected, and reprisals taken on his family. We did not see him again until it was all over.

We were sick with dismay. What Jumbo had related to us was a fiasco. The Americans should never have been expected to send men out of their battle plan to support such an initiative, unprepared, unknown, possibly a trap. If only Jumbo had told us we would have implored patience. We would never have let him advance such a harebrained scheme if he had any influence at all. And now twenty lives of the best kind of Germans had been thrown away.

We supposed that if their coup had succeeded, presumably by US intervention, they would have brought us into the act to prove their bona fides and strengthen their position. But now all we could do was to sit it out in our underground hide and hope for the best.

Then, one morning, Willy's elder son told us that the Americans had crossed the Danube. From then on he watched the main road running 200 yards away from the house and brought us frequent reports.

A band of prisoners from a concentration camp was visibly whipped along it by blackshirted SS. He saw one of the women bend as if to pick something up, a piece of something to eat. A guard shot her point blank with his rifle. She lay dead on the road while the miserable tribe went out of sight. The SS were evacuating and taking with them those of their victims they had not already murdered. A lorry full of German deserters was also driven by. Their hands were tied together. Willy's elder son followed from a distance and saw the SS hang them a kilometre away.

Next day, the Americans did not seem to be in any hurry to reach us, but as it wore on, German tanks began to pass southward, retreating on the road. The whine and rattle went on most of the day. Then came hours of silence, briefly shattered by low-flying aircraft, then the night. After years of trying to escape we were on the threshold of liberty. The Americans would almost certainly arrive next day and we would be free, provided nothing went wrong meanwhile. So far our luck had been unbroken. We had escaped from the camp with almost ludicrous ease, compared with all our previous efforts. We had been cleverly and quickly conducted to our hiding place by Jumbo and been well hidden and well fed by Willy. And now all we had to do was wait another few hours and our ambition for which we had worked through so many weary months would be simply and quietly realized. We slept at last, after hours of mingled excitement and hope.

At break of dawn, one of Willy's sons slipped in. He said that on no account should we raise the top of the shaft. A group of SS had dug themselves in on the edge of the road, evidently to fight a rearguard action when the Americans came. If we were discovered we would be shot out of hand and so would Willy and his family. 'For God's sake keep still', said the boy, and slipped away again.

The SS would certainly not surrender. We thought that, beaten back from the road, their survivors would resume position in Willy's and his neighbours' houses and in that case some might turn the car chassis on its side for cover and shooting, and our tunnel mouth might be discovered.

185

Doleful memories of failed escapes came crowding back. The mass of experience showed, illogically but unmistakably, that Kriegies nearly always got caught, and lived to dig another day. This time it could be different.

The battle started. We heard a distant machine-gun fire far away and soft like a cat purring, and a remote crackling of rifles, like twigs in a camp fire. Then quickly the volume increased and there was firing all around us. Bullets whistled busily through the air, smacking into the earth above us. We looked unhappily at each other. In our time we had been happy enough fighting in the air, hearing nothing. But this hole-and-corner business was unfamiliar.

We heard one particular gun firing. Then it came nearer. Soon it was firing immediately overhead. The firing continued for some time. Like the calm centre of a cyclone, we were apparently surrounded by it. It reached a crescendo. Then came an immense bang. Then silence for a few moments.

We heard running footsteps: a man clattered against the car, fell on to the floor of it. Then we heard someone fumbling with the trapdoor. We looked at each other. Manser drew his knife, stood under the entry, and switched off the light.

But our luck had held after all. As we listened to the trapdoor being lifted we heard the wonderful sound of tanks and tanks and still more tanks. A well-known husky German voice called down the tunnel. We switched the light on again, Manser replaced his knife, and Willy, wildly excited, cheering and a quarter drunk, came rushing down the tunnel.

In his eagerness to tell us the good news he had tripped over the car door and given us our fright. The Americans had arrived. The final bang had been a tank firing point blank at the SS trench which had been wiped out to a man.

The engagement had taken an hour and was described that night on the BBC as 'the stiffest resistance the Americans

186

have encountered since crossing the Rhine', which we could scarcely believe.

We crawled from the car, blinking in the bright sunshine. We had been nearly two weeks below the ground and were cold. The world was whirling with delight. After the meagre German transport we had seen on the roads in the last few months the American tanks and lorries and jeeps seemed as multitudinous as ants. We drank with Willy, had baths in his house, ate a bibulous lunch with the whole family. We exchanged addresses, gave him most of our cigarettes and prepared to leave.

Then Willy thought of photographs. We took turns to climb down into our shelter for Willy to photograph each of us climbing up and out again.

'What are you guys up to?' asked an American Captain, leaning from an American armoured car.

'We are British officers', we replied, not exactly to the point but sufficiently for him to drive us back to the camp.

As we neared it, we saw hundreds of Germans being marched away into the captivity we had left, and men in Allied uniform, not disarmed helpless prisoners, but wearing steel helmets, carrying pistols and Tommy guns and looking like conquerors.

We were hilariously received at the gate by Jack, an American Captain on guard there, whom we had left at the Centre Compound when we moved to Belaria, over a year ago. If he had a noticeable feature, it was a habit of joking at us about our escaping games, especially when they failed. We had not been blessed in knowing his transplant to Moosburg. And there he stood, idiot as ever, convulsed at his idea that we had failed again. With rather curt nods we did not disenchant him, and passed in.

We learned that a certain amount of shooting had taken place around the camp in the morning, effectively finished when a US tank drove right through the wire. Then there was wild excitement and an immediate change of flags. A dreadfully vulgar, though mercifully short, speech by General Patton who drove in and out, and every man began thinking how quickly he could get home.

23

Up And Away

Past the gate, my first memory now is of Jumbo in his corporal's uniform, sloppy and without his cap, standing lonely before an oval of US and British officers who were seated and confronting him. The issue was whether he should be marched away as a prisoner like nearly all the other German personnel of our camp, or set free to go home. He had called on Manser and me. We arrived and described what he had done for us, and what we knew of his anti-Nazi activities. We laid it on strongly. There was our real friendship for him. He was released the same day. Manser and I walked him out of the camp into his proper identity, Doktor Heil, a publisher.

The next memory is of opening our eyes in the mauve end of night, shivering from drizzle through a feeble tarpaulin. It was less than twenty-four hours since the American tank had obligingly blown the SS crew to small pieces almost above our heads. It seemed longer. After a scruffy slice of bread and thin German jam, we scanned the camp. It looked as filthy as ever, but bathed in the invisible light of being free at last. To savour it, we scrambled over the perimeter wire. American troops were manning the wire, we couldn't think why, and on we went into the dark trees as the thin rain stopped.

How nice it was to be quiet, away from the uncouth

throng. We set a fast march and as the shadows lightened, laughing like school boys, we tried to outpace each other, almost running, our legs wet to the thighs. It seemed time for things to stop being serious. Eventually we loped into a small clearing and stood still for breath. At the same moment, from the opposite side, a tall man emerged in a long, black coat. He stopped abruptly on seeing us as if to jump back into the wood. Our 'Good Morning' in German kept him still. Two steps toward him showed a drenched SS officer's kit, the insignia removed, crumpled from sleeping under the trees. He switched his head to and fro, saw there was no one else about and said 'Please let me be. I'm trying to get home. We are poor devils now'.

He spoke a gentleman's German and we had no thought of troubling him. Just then the sun stabbed through some leaves behind him, a minute or two of chat and we offered him a cigarette. He took it so eagerly that we gave him a packet we had. In return he said: 'Watch out here. There are all kinds of people about. Some might use their pistols if they see you here alone'. Oddly, we had not thought of that. We wished him well, turned about and beat a smart retreat through the gold and green on the tree trunks, still wet above us, back over the wire and into the universal debate of how long would it be to get home.

The answer was less than happy. Orders streamed in from high command, American and British, for the hundreds of thousands of ex-Kriegies throughout the camps in West Germany to stay put and wait for suitable military transport to take them to airfields in quite distant parts. In the East, under Russian command, repatriation was to take longer still. In the West the orders made dreary but evident sense, and they were generally obeyed.

But as Manser and I had already escaped from the camp more than half a month ago, when it was held by the Germans, we saw no reason to go back to it now. So we gathered what little paraphernalia we had left, including an impatient squadron leader, whose conscience we thought was his own concern, especially as he had a revolver and some ammunition which could become useful on the way. So in the early afternoon it was into the trees again, walking and no turning back; how long would it be to get home?

After an hour, a main road cut across the forest and we with it, thanks to a US military truck. This looked like getting home at speed, and so it continued. In the evening, the young American officers were hospitable to us in their mess. Theirs was standard military food, but delicious to us who could only eat half our helpings as our stomachs had tightened over the years. I still see the glance we each gave the others when one of the Americans left an apricot untouched on his plate. How could he? Proper beds, how lovely to sleep!

Morning, and they gave us north-bound seats along main roads, with burned-out tanks and smashed cars lying along the green hill sides and trees. Midday, and we drove through Ludwigshafen-Mannheim, a major town recently and hugely damaged by daylight and night bombers, buildings tumbled to great stones, and everywhere the sickening stench of thousands of corpses beneath them. Night, and we stopped in a small town with a Royal Air Force outpost. For some reason, now forgotten, they asked us to talk in German to the mayor, and I remember how strange and unpleasant the fear was on both his face and his wife's.

A couple of days more and we were in Brussels, a day of officialdom on who was who. Soon after the next dawn, the three of us sat with hundreds of other ex-Kriegies around a large military airfield where every few minutes a Dakota would be loaded and take off for England. But two hours sitting on the grass and, despite my fancied unyielding force through this last lap, I was suddenly dreadfully sick and taken away to the sick bay. Good-bye home that evening, good-bye Manser for a year or two (he was going back to Cambridge), good-bye Squadron Leader who never shot his pistol, good-bye food of any kind until the day after tomorrow.

How dreary were the next twenty-four hours. And then, what a marvel! I heard that my own squadron, 139 (Jamaica) Squadron, was quartered on the Station. I called and, behold, my friend Ted Sismore, still the observer of magnificent Reggie Reynolds, the leader of our squadron, came to see me after lunch before the doctor let me go at the end of the afternoon. The first thing Ted said was: 'You know, it shook the whole squadron when you went down'. For a moment, I had to fight to keep back tears.

Then he unfolded the story of our lovely Mosquitos since I left them. George Hodder, my navigator, was alive and attached to a Group Captain elsewhere. Ted told me who had been killed that I knew from our squadron and 105, its sister. For the eleven months since I went missing until D-Day, they had done much the same work. I was spellbound by the different strategies from Normandy until the erosion of Berlin, and nostalgic for the years I had lost.

I joined the mess in the evening in my appalling uniform. Reynolds was away and there were few faces I now knew. All the same, it soon turned into a roaring, come-home party, so much so that later in the night it became clear to everyone that as I had flown out in my Mosquito and bombed Duisburg, I should fly back in a Mosquito next morning to complete the operation.

From that moment life quickened like a dream. Soon after sunrise, what kindness, the bomb sights were dismantled so that I could lie on my front in the nose of the plane, an almost unheard of third passenger. It was a blue sky day. Flying Officer Howard, who had joined our squadron after I became a prisoner, and his navigator had been given a north-west course over the North Sea and then westward to land at Harwell, south of Oxford, the nearest to home.

His take-off felt like rising to heaven. Through perspex as clear as glass I gazed over the spring fields and trees of Belgium and thought of home and family. Then, ahead, the blue green sea. After we left the coast Howard swept deep down, we tore low level across the water like the old days, and the sunlight danced on the waves all the way.

Then up and over the English coast, cruising for a while, and a caress of a touch-down. Howard stopped off the runway and bustled me through the hatch to drop to the ground. I began to walk in England, the plane up and away, the story over.

we came
to no terms
with captivity,
dug tunnels,
climbed wire,
in the end
won free.